A JOURNEY to MECCA and LONDON

A JOURNEY to MECCA and LONDON

The Travels of an Indian Muslim Woman, 1909–1910

BEGUM SARBULAND JUNG

Translated and edited by

DANIEL MAJCHROWICZ

INDIANA UNIVERSITY PRESS

This book is a publication of

Indiana University Press
Herman B Wells Library 350
1320 East 10th Street
Bloomington, Indiana 47405 USA

iupress.org

First printing 2025

Cataloging information is available from the Library of Congress.

ISBN 978-0-253-07164-4 (hardback)
ISBN 978-0-253-07169-9 (paperback)
ISBN 978-0-253-07170-5 (ebook)

CONTENTS

ACKNOWLEDGMENTS

I FIRST READ BEGUM SARBULAND JUNG'S travel account, *The World through a Woman's Eyes*, in 2012. I encountered it in a dusty but peaceful corner of a university library in New Delhi. At the time, I was conducting research for a book on the history of the travelogue genre in South Asia. I had never heard of Begum Sarbuland Jung, but I was instantly captivated by her book's bold title. As I thumbed through the volume, I began to suspect that I had stumbled on a hidden gem. My suspicions were confirmed by the date of the travel diary's composition. Written in 1909–10, this was in fact one of the first full-length travel accounts ever written by a Muslim woman in modern India. Over the years, as my research on the travelogue genre progressed, I became increasingly convinced that Begum Sarbuland's account was a literary and historical landmark that was long overdue for translation and study. The present volume intends to meet that need and introduce Begum Sarbuland and her insights on Islam and the lives of Muslim women to contemporary English-speaking audiences. It also brings Begum Sarbuland's voice to an ongoing scholarly conversation about women's social and literary contributions to Muslim societies in early twentieth-century India.

When I first read her travelogue, there was no historical information available about Begum Sarbuland whatsoever. Frankly, beyond the fact that she had married a prominent lawyer and judge from northern India, I had almost no idea who she was. Thankfully, that is no longer the case.

Researching Begum Sarbuland and her fascinating life has been a long and exciting journey, one that I could never have completed on my own. I begin these acknowledgments by expressing my gratitude to all those who encouraged and supported me through the process of writing this book and translating Begum Sarbuland's voice for readers today.

Above all, I am grateful for the enthusiasm and patience I have been shown by Begum Sarbuland's surviving relatives. Their contributions to the final form of this book cannot be overstated. When I began my research, I found hints of Begum Sarbuland across the colonial archive, in English and in Urdu, but struggled to weave them into a coherent narrative and produce a meaningful picture of her life. Even the most basic details one might wish for in a biography, things like her birth and death dates, eluded me. Exasperated, I finally stepped out of the library and turned my focus toward locating a relative of hers who might be able to give me some leads. Early clues came from the website of the late Jamal Khwaja, who mentioned in an online article about his family history that Begum Sarbuland had been his grandmother. This reference gave me a leg to stand on. I wrote to Jamal, who responded enthusiastically but sadly passed away before we could continue our correspondence.

A second breakthrough came when I stumbled on a Facebook post that led me to Begum Sarbuland's great-nephew, Mirza Jahandar Baig, who happened to live in Chicago, just a few miles down the road from me. Jahandar enthusiastically shared data on his family ancestry and provided me with helpful photographs and other important leads. Soon, the begum's bare-boned biography began to take on flesh. In addition to treating me to Hyderabadi *kheema* and Chicken 65 on Devon Street, Jahandar also connected me with Begum Sarbuland's grandson, Najeeb Jung. Without Najeeb's assistance this book would be much the poorer. He spent hours sharing memories of his grandmother with me, once over burgers in Hoboken, later over kebabs at the Delhi Golf Club, and almost weekly via WhatsApp. For nearly two years, he has shared pictures, references, and reminiscences in the service of this project. He also helped me collect photographs from family members and connected me to other surviving relatives who grew up with Begum Sarbuland.

The first person he connected me with was Mrs. Shakira Sultana, Begum Sarbuland's granddaughter. She too graciously gave me a lengthy interview, shared relevant photographs, and has allowed me to pester her repeatedly with inquiries small and large. Today, my phone is brimming with messages and voice notes from her containing eyewitness information and familial insight on Begum Sarbuland's life in preindependence Hyderabad. At my behest, Shakira *apa* also cracked open and pored through Begum Sarbuland's private notebooks, which she has preserved, to check for any detail I might need.

I am incredibly grateful for and indebted to Abid Ilahi, who was surely a professional historian in a former life. This book has been immeasurably enriched by his enthusiasm for the project and the countless hours he has invested in its success. Abid has gone well out of his way to provide insights, share memories, track down information, and locate, scan, and upload many of the photographs that adorn these pages. In the final stages of this project, with a publication deadline looming, Abid took on this work as though it were his full-time job. He read drafts of the biography, cross-checked many of the details with his relatives, helped me untangle family relations, and carefully calculated birth and death dates. It was through his meticulous detective work alone that nearly all the individuals who appear in the photographs reproduced in this book have been identified.

Like a good historian, Abid was adamant that I consult family accounts of Begum Sarbuland from a variety of perspectives. Through his good offices, I did. Saberah Malik shared with me the stories that her mother, Taj Begum, would tell her about her grandmother. She also detailed Begum Sarbuland's interest in fashion and photographed a dress that Begum Sarbuland designed for the Delhi Durbar of 1911. In the final weeks of the project, I learned that one of Begum Sarbuland's grandchildren lived just two blocks away from my office in Evanston. I thank Muazzam Ilahi for immediately agreeing to meet me for coffee and share his own memories and reflections of his grandmother. I am also thankful to Begum Sarbuland's granddaughters, Mrs. Sehar Mansoor and Dr. Zehra Jung, for their insights and for allowing me to use images from their private collections. Jawahar Khwaja, who maintains a phenomenal collection of images of his family, generously shared many of these with me. Asma Jung cheerfully scanned files, uploaded images, and patiently fielded my many queries.

For me, speaking with Begum Sarbuland's family had the effect of bringing this project to life. When I began, I believed that I was reading words from the past, written by a woman who had long since left this earth. It was a purely historical study. Now, though, I realize that even seventy years after her demise, Begum Sarbuland and her legacy persist, and not just in memories. Begum Sarbuland dedicated herself to her children and grandchildren. By speaking to so many individuals who benefited directly or indirectly from her teaching, guidance, and determination, Begum

Sarbuland has, for me, been transformed from an ephemeral historical figure to someone whose legacy remains immediately tangible.

Beyond Begum Sarbuland's descendants, I am also indebted to my professional mentors, friends, and family for their support in making this book possible. The true genesis of this book came in 2015, when I began work on a volume on travel writing by Muslim women from around the world (*Three Centuries of Travel Writing by Muslim Women*, published by Indiana University Press) that I coedited with Sunil Sharma and Siobhan Lambert-Hurley. It was for that book that I first wrote about Begum Sarbuland and translated portions of her travelogue. Both Sunil and Siobhan have published extensively on women's travel writing in Urdu, and they have been invaluable mentors for my work in this field. Both provided encouragement and feedback and suggested secondary source material. I also benefited from discussions of my work on this text at several conference presentations and lectures, including at the Annual Conference on South Asia at Madison in 2016, the Berkshire Conference on the History of Women, Genders, and Sexualities in 2017, the University of Chicago's Workshop of the Center for the Study of Gender and Sexuality the same year, Delhi University in 2022, and the University of Lausanne in 2023.

At Northwestern, I have had many conversations about this book with my colleague Laura Brueck, a champion for the value of academic translation. Her encouragement played a major role in bringing this project to fruition. I have also benefited from conversations from many other colleagues, friends, and mentors, including Gail Minault, Barbara Metcalf, Francesca Orsini, C.M. Naim, Rajeev Kinra, Melissa MacCauley, Sylvia Vatuk, Blain Auer, Nishat Zaidi, Roanne Kantor, Andrew McDowell, Timsal Masud, Arafat Razzaque, Melina Gravier, David Boyk, Katherine Merriman, Haris Qadeer, Asiya Alam, and Sabeena Sheikh. Ragy Mikhaeel helped me decipher some of the more esoteric renderings of Arabic conversation included in the diary. David Boyk and Andrew McDowell have talked through every aspect of this project with me. Both have read and commented on much of its contents. Sumaira Nawaz has talked endlessly about this work with me, allowing me to bounce ideas off her and always giving the wisest council. Abdurrashid helped me understand difficult passages and patiently explained unfamiliar familial terminology.

Two brilliant and motivated undergraduate students at Northwestern, Hajra Malik and Hamnah Malik, read the original translation, provided feedback, helped me check tricky sections against the original Urdu, and assisted me with various research tasks. My collaboration with both was facilitated by funding from Northwestern's wonderful Undergraduate Research Assistant Program. Even after the grant finished, Hamnah stayed on, transforming from a research assistant to an excellent editor. She has provided critical editorial advice. Anusha Tomar indulged in debates about the nuance and meaning of many passages in the travelogue and suggested ways to improve my writing. Rija Ahmed spent countless hours cheerfully checking every doubt I had about the meanings of individual lines. After submitting the initial manuscript to Indiana University Press (IUP), two anonymous readers gave me several pages' worth of insightful feedback that helped bring this book to its final form. Bethany Mowry, Sophia Hebert, Lesley Bolton, and Samantha Heffner at IUP have been the best editors and project managers that one could hope for, and I feel incredibly lucky to have been able to work with them for a second time. Matthew Perez provided careful copyediting.

Special thanks are due to the Alice Kaplan Institute for the Humanities at Northwestern University. The institute provided a generous subvention that allowed me to include a rich collection of high-quality images with this book. James Nicholson, Qaisra Khan, and Nahla Nassar assisted me in locating and preparing images of the Hjiaz Railway included in this book. I thank the Khalili Family Trust for allowing me to include these rare photographs.

Finally, I thank the librarians and staff at Jamia Hamdard for their hard work to preserve Urdu's literary heritage and history. They allowed me to access their collections, which includes one of the last remaining copies of Begum Sarbuland's travel account. It was at Jamia Hamdard that I first read Begum Sarbuland, and thus, it is through their efforts to preserve Urdu's literary heritage that this project was ultimately put into motion.

NOTE ON TRANSLATION

Begum Sarbuland's travel account began its life in 1909 as a private travel journal. It was only published decades later with light editing. The production quality of the print version was low and the book contains many transcription errors. I have corrected obvious mistakes in my translation but generally noted their occurrence. In my translation, I have also tried to approximate the journal's easy informality for an English-speaking audience. To this end, I have particularly shortened or removed lengthy honorifics that are intuitive in Urdu but cumbersome in English, though I have left them in place in section headings and where they form a logical unit, as in "Nawab sahib." The journal also contains a moderate number of English words (such as "lunch" or "breakfast") that Begum Sarbuland presumably used in her daily life, even when they were unfamiliar for general Urdu readers. She retained these English words in her published diary but often provided parenthetical glosses. These glosses are of course unnecessary for English-speaking readers, and I have excluded them here. On some occasions, I indicate the use of English vocabulary with quotation marks. Another idiosyncrasy of the book is its shifting use of tense; presumably, some diary entries were written as reflections on days past, while other entries describe events as she experienced them. At other times, even when clearly narrating events in the past, she switches between using the past and the present. In English, these rapid shifts are jarring. Thus, even as I have attempted to retain the book's meandering temporality, I have, on occasion, regularized her use of tense where confusion might arise. These slight modifications do not affect the meaning or the feeling of the text, but they do make the translation more readable. The original text is divided into 118 sections. I have retained the original section headings but divided them into ten chapters. Unfamiliar terms that appear often in the diary are defined in the glossary. This book uses a consistent transliteration scheme for words and proper names in Urdu, a simplified version of the system advocated by the *Journal of Urdu Studies*. There are a few exceptions: when words have been adopted into English, like "purdah," "nawab," or "begum," I use the more familiar spellings. In keeping with a longstanding family practice, I have rendered the honorific title (and later surname) "Jang" as "Jung."

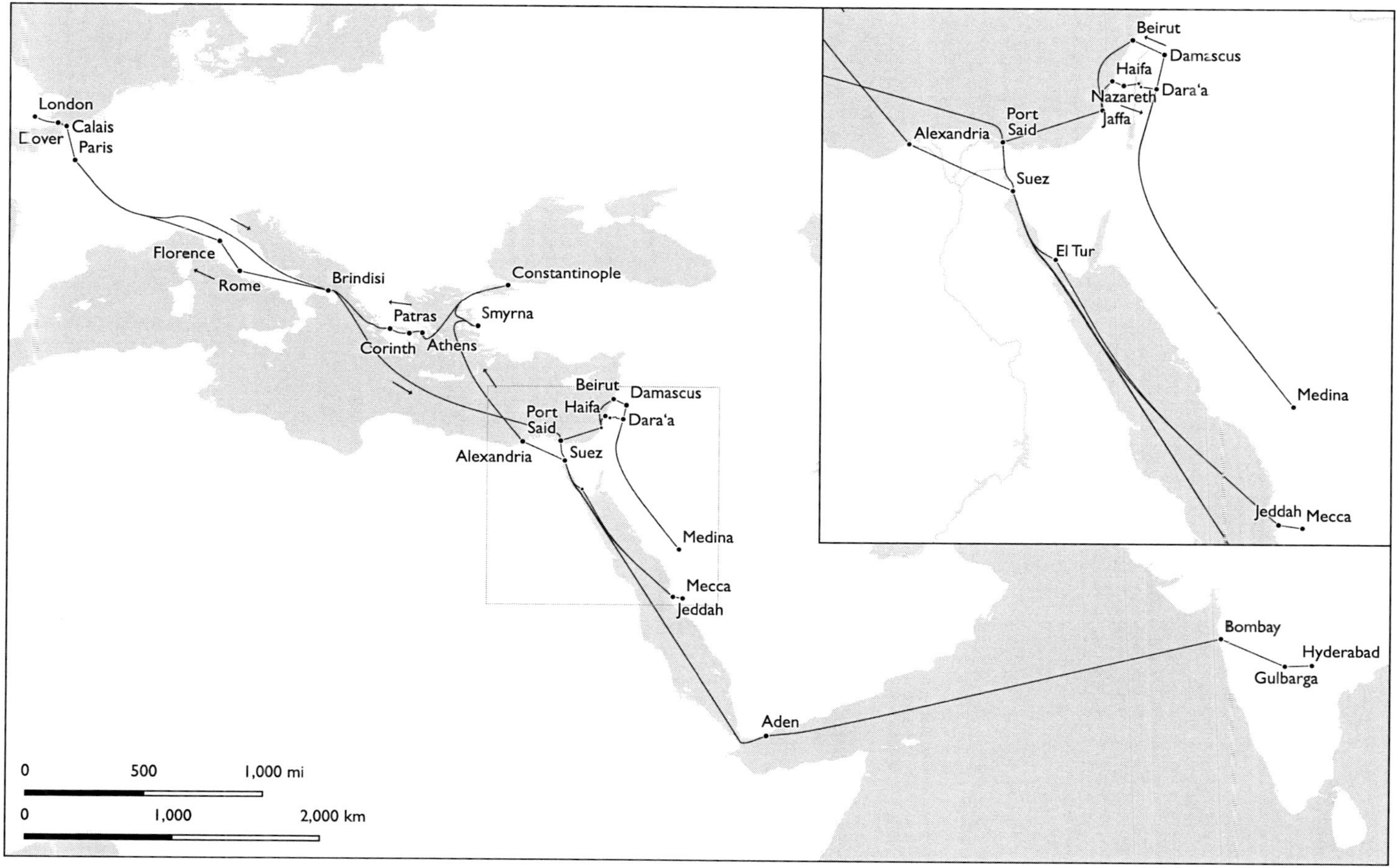

Map The route traveled by the Sarbulands from Hyderabad, India, to London and back.

A JOURNEY to MECCA and LONDON

INTRODUCTION

On an early summer evening in 1910, an Indian gentleman strode across the platform of Bombay's Victoria Terminus railway station. Heads turned as he went. And why not? Nawab Muhammad Hamidullah Khan Sarbuland Jung (1864–1930) was, after all, the influential chief justice of the Hyderabad Supreme Court and the poster child of Indian Muslim colonial modernity (fig. 0.1). Hamidullah Khan's prominent role as head of the judiciary in one of the most powerful and wealthy kingdoms in pre-independence India made him an instantly recognizable figure as he boarded his first-class train carriage. However, few of his fellow passengers would have been able to recognize, much less name, the burqa-clad woman who entered the train along with him.

On that evening in Bombay, Nawab Sarbuland Jung was on his way home to Hyderabad after having spent four months traveling through the Middle East and Europe. It had been a packed itinerary. He had toured Damascus, Medina, Athens, and Paris. He had performed the hajj in Mecca and met the Ottoman sultan in Constantinople. In London he had visited the Prince and Princess of Wales and spent some tender moments with his young son, who was studying at a boarding school nearby. Though he was not traveling on official business, his public renown meant that his journey was widely followed in European, Ottoman, and Indian newspapers, at least one of which printed a large picture of him in Mecca, decked out in

Fig. 0.1 A photograph of Nawab Hamidullah Khan Sarbuland Jung, probably taken in late 1909. *Source: Daniel Majchrowicz.*

Arabian garb.[1] Now he was headed home to a hero's welcome. As many as five hundred friends and well-wishers were waiting at Hyderabad Station to celebrate his return under the careful gaze of the journalists dispatched to cover the event. Everyone, it seemed, had heard about the nawab's grand adventure.

There was one aspect of the journey, though, that few people knew about. Though the newspapers following his voyage made no mention of it, the truth was that Hamidullah had not made this trip alone. At every moment, he had been accompanied by his spouse, Akhtar un-Nisa Begum Sarbuland Jung (1876–1957). Begum Sarbuland too performed the hajj and toured Damascus and Paris. She also met the sultan in Constantinople and bantered with the Prince and Princess of Wales in London. And she was there in Bombay as well, though far less recognizable as she boarded the night train in a black Turkish burqa and face-concealing niqab. She had not worn this covering uniformly throughout the trip, for Begum Sarbuland adapted her veiling practices according to the social circumstances and her mood.[2] Still, like most elite Muslim women of her day, when in India, she preferred to maintain a more stringent level of purdah.

Begum Sarbuland's purdah complicated the logistics of her train journey from Bombay to Hyderabad. With a crush of supporters waiting for her husband on the platform, simply alighting from the train with him was impossible. Instead, Nawab Sarbuland suggested that she exit the train one station early. In fact, he had already wired ahead to ensure that some of Begum Sarbuland's female relatives and friends would be there to welcome her. All of this was in keeping with the gendered social practices of the upper classes in early twentieth-century Hyderabad. The state was famously conservative, and gender segregation and purdah were the norm among the city's Hindu and Muslim elite.[3] In this milieu, a wealthy noblewoman like Begum Sarbuland was expected to avoid the public gaze. As a man and a prominent government figure, Nawab Sarbuland was expected to court it. By the same token, while her husband's views about Europe were public knowledge, Begum Sarbuland's views remained under wraps. Unless you were a part of her family or a close member of her social circle, you would not have had the privilege of hearing her perspective.

One way Nawab Sarbuland shared his experiences and impressions of the journey was through the medium of travel writing, a literary genre that was all the rage in turn-of-the-century India.[4] In Hyderabad particularly, the travelogue was a major component of social and intellectual culture. Nawab Sarbuland took the responsibility to write about his travel seriously. He ultimately produced three full books and numerous newspaper articles in Urdu and English about this one trip. More than simply fulfilling a social expectation, his travelogues also let him demonstrate the breadth of his experience and his capabilities. His wife performed her own socially expected role by not publishing anything at all.

Though Indian women were frequent travelers, there were serious practical and social impediments that prevented them from sharing their reflections in print. For one, literacy among women in India was abysmally low: in 1901 fewer than 1 percent of women could read.[5] Moreover, the world of Indian print was overwhelmingly dominated by men. Women writers were often not taken seriously and, worse, risked social ostracism and even violence if they tried to barge their way in. Some women also opted not to publish their writing according to their own sense of propriety. This was probably the case with Begum Sarbuland, who had the resources and ability to publish a travel account, as a few of her peers were beginning to do, but initially decided not to.

Twenty-five years later, though, a series of factors led Begum Sarbuland to publish a written account of her journey after all. For one, society's views about women had changed. There was significantly more support for women's education, and a handful of Muslim women had already printed travel accounts. By the 1930s, putting one's name into print was not nearly so radical as it had been.[6] Begum Sarbuland's personal circumstances had changed as well. When she printed her travelogue, sometime around 1935, she was about sixty years old. All her children had grown up, and her husband had passed away. Whatever her inspiration—whether societal change, the liberation of widowhood, or simply her children's prodding—she decided to put her book before the public at last. Thanks to that decision, Urdu readers of her time gained a rare glimpse of what it had meant to travel the world as an elite Indian Muslim woman in the early twentieth century. Today, through the medium of translation, contemporary English-speaking readers can enter her world as well.

Begum Sarbuland's Travels

One night in June 1909, Begum Sarbuland was lying in bed when her husband Hamidullah turned to her and said, "Would you like to go on hajj this year?" Without hesitation, she answered that if God willed it, she would. This would be her first time abroad. She began to prepare for the trip almost immediately. The journey ended up being far more than a pilgrimage to the holy city of Mecca, as her husband had initially suggested. In addition to completing the hajj, which is typically considered incumbent on every Muslim who has the means to perform it, the couple also saw swaths of the Middle East and Europe. Even by today's standards, the journey was ambitious and lengthy.

The trip had three primary destinations: the Hejaz (particularly Mecca, Medina, and Jeddah) in today's Saudi Arabia, Constantinople (or Istanbul, as it is known today), and London. While all these locations were easily accessible from India by steamship, the couple avoided the beaten path and instead followed a highly idiosyncratic route between them that took them to corners of the Middle East and Europe rarely visited by Indian tourists (map 1). The reason for this seems to have been that Hamidullah had already traveled widely in the region. Their itinerary allowed him new experiences beyond the main steamship routes he knew so well. Thus, rather than going from Bombay directly to Jeddah, the port most used by Indian pilgrims to Mecca, the Sarbulands first sailed to Haifa, in Palestine.

Hamidullah was eager to travel on the newly constructed Hejaz Railway, an Ottoman-built railway line that ran from Damascus to Medina, and Haifa made for easy access to the line.[7] From Haifa, they first traveled east to the holy Christian city of Nazareth. (Most Indian Muslims visiting Palestine would not have skipped Jerusalem in this way.) Then, they took the train to Medina. Typically, travelers would proceed from Medina directly to Mecca by camel caravan, but, in another unexpected twist, the Sarbulands instead rode the train northward to Damascus. They stayed there for a single day before continuing to Beirut and boarding a series of steamers to Jeddah. From Jeddah, they went by camel to Mecca to perform the hajj. Until this time, they had been accompanied by a single female servant, Amina bi, who looked after them, cooked their meals, and guarded their luggage. In Mecca, this trio was joined by a large party from Hyderabad and Bombay that included several family members and life-long family servants. Among the latter group was the family's wetnurse, who had brought the couple's infant son, Halimullah, along!

After performing the hajj, the larger party returned to Hyderabad with Amina bi while the Sarbulands continued their trip alone. From Jeddah, they first boarded a ship to Constantinople but abandoned the vessel when they learned that, due to a plague outbreak in Jeddah, all ships coming from the city would be subject to quarantine in mainland Turkey. The couple craftily skirted the Ottoman cordon sanitaire by disembarking at Suez and taking a train to Alexandria (skipping Cairo!). Sailing from plague-free Egypt directly to Constantinople spared them the ordeal of an indefinite quarantine.

In Constantinople Hamidullah had some unnamed business that meant he was frequently off on his own. Begum Sarbuland was left to her own devices. Uncertain of her ability to navigate such a large and unfamiliar city, Begum Sarbuland opted to spend much of her time in the hotel, which also happened to be the finest luxury European-style accommodation in the city. Nevertheless, she did manage to tour the city's main sites when her husband was able to accompany her. She also attended several parties and even met the Ottoman sultan and Muslim caliph Mehmed V at his lavish European-style palace on the Bosporus. Despite facing significant cultural and linguistic barriers, during these social events Begum Sarbuland was thrilled to meet Muslim women from the Turkish aristocracy, with whom she sought affinities based on a shared religion and social class.

From Constantinople the couple traveled to Greece and then Italy and France. By this time Begum Sarbuland seems to have been very tired of the journey. Anyone who has ever made a touristic trip to Europe will commiserate with her sheer exhaustion at visiting one museum and garden after another. At one relatable moment, she begs her husband to spare her the torture of yet another Italian palazzo. The couple strikes a deal: she will continue the tour, but only in a convertible automobile. This lack of interest in mainland Europe was not purely a symptom of fatigue and sensory overload, though. She found Europe underwhelming compared with the emotional excitement of the holy lands and the sociopolitical pull of Constantinople, the heart of Muslim modernity and a beacon for Indian Muslims who yearned to see Islam regain its status as a religion of state in India and elsewhere in the colonized world. (In his travelogue of the city, Hamidullah called it "the Muslim capital.")[8] Begum Sarbuland identified strongly with her faith, on a political as well as a spiritual level, and was above all eager to understand contemporary Muslim life outside India and to meet her fellow Muslims in their homes and on the streets. By comparison, the glories of ancient Rome and contemporary Athens held much less appeal. Paris, it must be said, was somewhat more bearable to her, largely because the couple's attention had shifted from museum hopping to boutique shopping.

Begum Sarbuland's verve returned in full form when the pair reached London, a city she had heard about since childhood and whose language she spoke well. British power in India was then near its zenith, and London was a veritable Mecca for wealthy and up-and-coming Indians. It was a city filled with Begum Sarbuland's friends and family, many of whom were living, studying, or working there, and she fit right in. Her first priority was to spend time with her family: with her eight-year-old son, who was attending a boarding school in the countryside; her brother, who was studying medicine; and Hamidullah's nephew and their future son-in-law, a law student. (Plate 7 shows a studio portrait of the group together in London.) Nevertheless, she did make time to meet with other people in the city, including the future king and queen of Britain (the soon-to-be King George V and Queen Mary), whom she knew from their visit to Hyderabad a few years before.

From London, the couple set sail for Hyderabad. Begum Sarbuland was now running very low on steam. She was plainly relieved to be going home but also upset to leave her son behind. From London, she and her husband traveled to Brindisi and then to Bombay. Before her arrival in India, Begum Sarbuland resumed wearing the burqa that she seemingly donned only sporadically in Europe. Her arrival in Hyderabad, where she alighted from the train at a secondary station, was quiet but not anonymous. She was met on the platform by her mother and a few friends. The welcoming party immediately noticed that the long trip had invigorated her: "When I flipped my veil back, they saw that my face had been changed by the blessings of my pilgrimage to the holy places and by the European climate. . . . This is the light that God has placed upon me since my visit to the shrine of his Beloved Prophet."

Though so much of this trip was novel for Begum Sarbuland, she largely remained within relatively familiar social spaces. The Sarbulands' wealth allowed them to travel in enviable comfort. Their movements were largely coordinated by European tour operators, and they typically stayed in European-owned and -run hotels—except on one occasion in Damascus, when Begum Sarbuland's desire for Islamic unity inspired her to try out a "Muslim hotel," a decision she immediately regretted. Their modes of transport included steamships, trains, carriages, donkey carts, and camel litters. Whenever possible, though, they went first class. Moreover, they seemingly ate mostly Western food in five-star establishments or had their servant prepare meals for them, except in Arabia, where they had access to housing provided by the state of Hyderabad. Unlike some Indian Muslim travelers, neither evinced an aversion to eating non-halal food, and on the occasions when they had to eat continental European fare, Begum Sarbuland lamented fiercely the lack of puddings and other British "delicacies."

On some occasions, though, the Sarbulands' wealth could not guarantee their comfort. In the Middle East, the couple had largely relied on tour operators from Thomas Cook, a British company that specialized in facilitating touristic travel for Europeans (though it also had a history of organizing transport for the pilgrimage from India). Despite this, Begum Sarbuland's sense of curiosity meant that she often ventured out to chart her own course without her husband or a Cook representative as an intermediary. In these adventures, she often encountered cultural differences that she was forced to navigate alone. These passages are where her travelogue truly shines. One of the book's rawest moments takes place in the bathhouses of Damascus, where Begum Sarbuland was forced to completely abandon her own sense of personal space. By opting to stay in a "Muslim hotel," which did not have a full bathroom, she was compelled to bathe in the public hammam. With almost no ability to communicate inside, she begrudgingly abandoned her inhibitions and allowed the attendants to undress and wash her, as was customary. This was a new experience for her: Indian women of her class were deeply resistant to being seen naked, even by other women.[9] Ultimately, though, she came around to the experience and found herself wanting to visit a hammam once again, "in part because it gave relief to my body, but also because it was one of the more memorable experiences of our trip."

From Private Diary to Published Book

After Begum Sarbuland completed her handwritten travel diary in 1910, the volume remained within the family until around 1935, when she had it published with light editing. The present book is an English translation of that 1935 print. The diary's unique publishing history—a text meant only for close family and friends being published in approximately its original format twenty-five years later—is much of the reason that it is so remarkable and relevant to readers today. Had Begum Sarbuland chosen to publish a travel account immediately after her journey, as her husband had, it would surely have read very differently. Given her social position and the sensitivities surrounding Muslim women's writing in early twentieth-century India, such a book would have carefully adhered to the gendered standards of Urdu print. What is more, with her husband then at the peak of his career, its contents would have remained circumspect and studiously politically correct (as were her husband's writings). The result would have been far more restrained; the book would likely never have included such a candid discussion of being hand-washed in a Levantine hammam or of slipping a train conductor a bribe against her husband's wishes.

By reading the 1935 version, printed when she was already a widow, we gain access to her more intimate reflections, experiences, and emotions. We also encounter the thoughts and feelings of an elusive life, the type of figure that rarely appears directly in Indian or Islamic history except as refracted through men's depictions or as represented in the careful social codes that many Indian women of the period used in their public writing. As Siobhan Lambert-Hurley notes, "in women's writing [from colonial South Asia] especially, the self may be elusive. More often than not, in South Asian autobiography the story of the self is one of self-deprecation over self-aggrandizement, with individual lives told in relational terms."[10] In private writing, though, this was less likely to be the case.

Begum Sarbuland herself realized that her book offered a rare opportunity for readers to see the world "through a woman's eyes." As this was her first trip abroad, her account glimmers with excitement, curiosity, and wonder. Through it, we learn what she felt as she spent her first night at sea on a steamship. We follow along as she observes with fascination the diversity of Muslim societies outside India's borders and attempts to negotiate these unfamiliar and sometimes disconcerting spaces. And we are horrified, and yet impressed, by the calm magnanimity with which she confronts casual European racism, as when a British woman forcibly removes her veil in public.

The book also reveals how she reflected on her own beliefs, practices, and preferences in light of these experiences. What did it mean, as a Muslim woman from colonial India, to travel through the Ottoman Empire, to meet the Muslim caliph? Was it unsettling to encounter Muslim women abroad whose languages, practices, clothing, and even beliefs were so different from her own? And above all—to use her own words—what did it mean to be a Muslim woman in a time of rising Westernization? What was it like to make her way through a colonial world that was not designed for practicing Muslims like her, a world unwilling to fully accept her? It was precisely these reflections and insights that she wished to share with her readers.

Her account also sheds light on the mundane: her everyday routines, her travel logistics, the nature of her relationship with her husband. Her writing obliterates tired stereotypes of oppressed, meek, or timid Muslim women imprisoned behind the four walls of the home by jealous male guardians. For example, while the couple are in Mina, just outside of Mecca, Begum Sarbuland suddenly decides that she would like to go back to the city to visit the Kaaba before sunset. Her husband does not want to join her, so she gathers a small escort, mounts a donkey, and rides off. When she reaches Mecca, she abandons her escort but finds the sacred space around the Kaaba thronged with pilgrims. She wonders how she will ever make her way through such a dense crowd alone. Suddenly, she is approached by an African man who asks her, in Arabic, "Oh, my lady, do you need a kiss?" She answers without hesitation: "Yes, I do."

The man takes her hand and, maneuvering through the crush of people, bears her to the *hajar aswad*, the revered black stone set into the Kaaba's outer wall. She is shocked by this occurrence—it requires great luck and a fair bit of physical strength and tenacity to reach the stone in pilgrimage season. With no time to reflect on the immensity of this miracle, she kisses the stone repeatedly as he shields her body with his own to protect her from the pressing crowds. He then leads her back to safety. The pair talk and pray together (another miraculous occurrence, since they do not share a language), and then she remounts her donkey and returns to Mina. Clearly, Begum Sarbuland felt free to chart her own course, to make her own decisions, and to go her own way. She often avoided unnecessary contact with unrelated men, but as this anecdote shows, her rules of engagement were varied, complex, and entirely of her own making.

Perhaps recognizing the power of her private travel diary and the value of its candid insight for her contemporaries, Begum Sarbuland did not compose a fresh, polished narrative when she finally decided to tell her story. Instead, she edited her old diary (probably adding section headings), wrote up a preface, and sent the manuscript to a printing press near her home in Old Delhi. She called the book *Dunya Aurat ki Nazar meñ: Mashriq o Maghrib ka Safarnama* (*The World through a Woman's Eyes: A Travelogue of East and West*) (plate 1).

This title foregrounded one of the great causes of her life: the search for an ideal and uncompromising Muslim womanhood that could thrive in a Western-dominated world without succumbing to it. She addressed the book directly to women readers, whom she called "my sisters." This language suggests that Begum Sarbuland believed she could offer a unique perspective for women readers, who had relatively few opportunities to see the world through the eyes of someone like them. The published book contains many grammatical errors and a few passages that seem repetitive or contradictory. This roughness, though, only serves to strengthen the visceral sense of encountering Begum Sarbuland's thoughts and feelings directly, as she wished for them to be experienced by her Muslim sisters.

Prominent Themes in *The World through a Woman's Eyes*

Begum Sarbuland's diary is among the earliest known travel accounts by a Muslim woman from anywhere in the world.[11] *The World through a Woman's Eyes* thus offers rare insight into a host of historical questions about colonialism, Islam, mobility, class, race, and gender. Some of those questions, many of which are vexing, are addressed below.

Travel with Servants

One of the most intriguing and ephemeral figures in the book is Begum Sarbuland's lifelong female servant, Amina bi. Having a servant accompany wealthy Indian travelers was not exceptional, but having a servant appear in travel writing was. At best, servants were mentioned in passing, as exemplified by Nawab Sarbuland's curt reference to Amina bi in his own travelogue. Yet in Begum Sarbuland's private travel diary, this close companion of hers appears again and again. Amina bi's presence, even when reflected through her employer's pen, make clear that her experiences abroad were worlds apart from Begum Sarbuland's. Though we never hear from Amina bi directly, attempting to reimagine the book's passages from her perspective allows some access, if speculative, into the experiences of underrepresented travelers like her.

Amina bi accompanied the couple through the first half of their trip. She was given a multitude of tasks to perform and was often left to navigate foreign spaces entirely alone and without the comforts of wealth, privilege, or cultural familiarity. On one occasion in Syria, Amina bi was instructed to go to the bazaar to buy a chicken to cook. Later, after settling into their first-class train compartment, the Sarbulands told Amina bi to go and find space for herself in the economy carriage. How did Amina bi, who likely had no linguistic ability beyond Urdu and probably no experience with foreign travel, manage to do all this? And how did she feel about it? One strains to discern her perspective as she followed the same itinerary as her employers, albeit from an entirely different vantage. Amina bi's enduring but muted presence stands as a reminder that this account was written by an elite, wealthy woman. Although Begum Sarbuland wants us to "see the world through a woman's eyes," Amina bi's silent presence reminds us that Begum Sarbuland's views were not representative of all Indian Muslim women of her day. At the same time, the intimacy of the two women's relationship as reflected by Amina bi's ubiquity complicates assumptions of rigid and insurmountable class or caste hierarchies in colonial India.[12]

Race, Language, and Begum Sarbuland's "Muslim World"

In her diary, Begum Sarbuland categorizes the people she meets in ways that confound our contemporary expectations. She wrote at a time when religious identity was moving to the forefront of Indian identity, as the idea of Muslims as a distinct community, both within India and globally, was crystallizing. Accordingly, Begum Sarbuland typically identified people first by their religion and only secondarily by their ethnicity, language, or nationality. Under this taxonomic system, only when an individual's religion was beyond doubt did she use these secondary classifications. For instance, in Palestine she described having dinner with "two Germans and a Jew." The Germans are identified by their nationality or language here because they were, presumably, Christian. "The Jew's" nationality or linguistic identity is never mentioned.

In the Middle East, Begum Sarbuland always identified those assumed to be Muslim by their ethnicity—Turk, Arab, Egyptian, and so on. By contrast, she labeled Middle Eastern Jews and Christians by their religion, as though they did not fully belong there. Despite frequently meeting local, Arabic-speaking Christians in the region, Begum Sarbuland seemingly believed that only Muslims could be Arab. In Mecca she resolved not to make any purchases at shops owned by "Christians" but to only patronize shops owned by "Arabs." Clearly, she felt that these were opposing terms. At the same time, her willingness to identify and celebrate Muslims abroad had its own limits. While she eagerly embraced Turkish and Arab Muslims as a part of her religious community, she typically did not extend that unity to African or Asian Muslims. She was surprised to find that the Somalis living in Aden could speak Arabic and offered no suggestion that they might be Muslims like her. In short, as was typical of her time, Begum Sarbuland's conceptions of who fully qualified as members of the Muslim community were fractured by race and ethnicity.[13] This is borne out succinctly in her anguish at seeing Christians in the Middle East who spoke Arabic but not English: "Even the Christians can speak Arabic, and yet I, a Muslim, cannot speak the Arabic language, the 'language of Muhammad,' as it is called. Some Christians can *only* speak Arabic and don't know any English at all."

Another telling example of the complex intertwining of race, religion, and belonging comes from her time in Constantinople. Here, she regularly met with a woman she calls a "memsahib." In Urdu, this term, which derives from the English word *madam*, is typically reserved for white, European women—and that is how Begum Sarbuland herself uses it in most cases. Yet this memsahib, a seamstress by profession, spoke fluent Arabic! The seamstress was not a memsahib in the sense of "European" at all but was actually an Egyptian Christian! For Begum Sarbuland, the woman's Christianity thus marked her as akin to a "European" rather than as an Egyptian or an indigenous citizen of a multiethnic Ottoman Empire and Middle Eastern world. Following this pattern, Begum Sarbuland almost always refers to non-Muslim Middle Eastern women as memsahibs and to Muslim women as "bibis" or "Arab bibis."[14]

The historian Nile Green has argued that the arrival of steam travel and the ability of Asian Muslim travelers to quickly move between Asia, the Middle East, and Europe presented them with a "bewildering experience of a collapsing former spacetime" that forced them to reimagine how they saw the world.[15] Port cities, he suggests, presented travelers with "confusing medleys of peoples from far and wide, whose ambiguous identities confused the conceptual borders between self and other."[16] Focusing on the travel writing of male intellectuals, Green compellingly shows how these jarring experiences led them to reimagine their conceptions of Muslim spacetime. Yet Begum Sarbuland demonstrates that the experience of steam travel did not always "reconfigure the most basic assumptions of all who experienced it."[17] What she saw did not cause her "to contemplate the very location and genealogy of Islam."[18] While her diary leaves no doubt that she saw a bewildering and unfamiliar world, she seems to have resisted refashioning her conception of who and what was Muslim.

In short, faced with the racial and religious diversity of the Middle East, Begum Sarbuland largely stuck to her guns—and to ideas of Muslimness then current in India. The same was true of Europe: she did not comment on the growing Muslim population of London at all. Her diary instead suggests that she believed in clear-cut and immovable distinctions between "East" and "West." She embraced the growing sense that Muslims belonged to a single community but limited who could belong to it, no matter what she saw abroad.[19] Thus, she did not accept the Somalis she encounters near Aden as Muslim but gushed with Islamic pride on seeing the "Arab" Muslims of Suez.[20] Similarly, Begum Sarbuland only noticed "Arab" and Indian women during her hajj and pilgrimage to Medina. She did not record encountering East and Southeast Asian or African Muslims, though these groups were present in huge numbers.[21] Rather than trigger a reevaluation of her most basic assumptions, her trip largely reaffirmed her views.

Piety and Religious Devotion

Begum Sarbuland's account is deeply devotional. That devotion is heartfelt but also shaped by her social standing. She and her husband were undeniably wealthy, and they got about in immense comfort. On this pilgrimage they generally stayed at the best hotels, traveled in first class, and shopped at the most expensive boutiques. In India, they had an entire train carriage to themselves, which could be attached to and removed from other trains according to their travel itinerary. Her journey was not always luxurious, and one can at times genuinely commiserate with her discomfort, as during the long train ride from Damascus to Medina, which she spent shivering in a doorless compartment fending off thieves intent on stealing her husband's golden shirt buttons. Nevertheless, the diary makes clear that she did not feel that her privilege was something to be downplayed. She neither flaunted her wealth nor pretended that worldly goods meant nothing to her. In Mecca, where Indian pilgrims often embrace the Prophetic ideal of poverty, the couple spent lavishly on devotional acts and Begum Sarbuland expressed a wish to return to the city again "with honor, prosperity, wealth, and health."

Next to her devotion to God and her reverence for the Kaaba, the diary prominently displays Begum Sarbuland's devotion to holy men, both living and dead. While traveling, Begum Sarbuland frequently wrote letters to her spiritual guide, Maulana Abd ul-Bari.[22] She had a close relationship with this leading Sufi, a man who would become a national political leader during the Khilafat movement. Yet even more than to living holy men, Begum Sarbuland and her husband directed their devotion to the deceased. The book opens with visits to several important Sufi shrines in Hyderabad, where Begum Sarbuland took her leave from the saints and asked for their guidance and protection during the trip. The couple's first stop after Hyderabad was also a Sufi shrine, that of the famous Chishti saint Khwaja Banda Nawaz Gesu Daraz in Aurangabad. In Istanbul they visited the shrine of Ayyub ul-Ansari, a companion of the Prophet, in Damascus the sacred shrines of the Prophet's family, and the graves of many other Islamic prophets besides. Begum Sarbuland did not visit these sites as a tourist but as a devotee. Her prose gushes with emotion as she comes into the physical presence of these revered figures. Of the burial site of the severed head of Imam Husain, the Prophet's grandson, she wrote: "The moment we arrived, my heart was overwhelmed by a strange sensation that I am incapable of describing. Nawab sahib said, 'Just look at the effect that this place has on our hearts.' 'Yes,' I replied, as we both began to weep. I wished I could stay there forever."

The true center of their devotion, though, and undoubtedly the most important site that Begum Sarbuland visited during her trip, was the shrine of the Prophet Muhammad in Medina. The Prophet was, and largely remains, at the core of Indian Muslim pietistic devotion, and Begum Sarbuland's passion for it is more palpable than anything else in the book. For most Indian pilgrims of her time, the obligatory visit to Mecca evoked strong emotions, but it was Medina and the Prophet's tomb that was most highly anticipated. In her 1935 preface, Begum Sarbuland reiterated this devotion: "My soul grows restless when I recall the lattice that surrounds the tomb of the Ruler of Medina [the Prophet Muhammad], when visions of its green dome appear before my eyes. My heart yearns to travel to the Land of the Beloved, to return to the soil of Medina and press it to my eyes. To let everything disappear but my Lord's tomb, its lattice, and I." This expression of an all-enveloping devotion to the Prophet, while not surprising, is important to emphasize. Early twentieth century India witnessed the popularization of modernist, reformist interpretations of Islamic doctrine. This form of Islam deplored devotion to the deceased and derided any practice that might be considered an affront to strict monotheism. But the Sarbulands seem untouched by these ideas, in 1910 as much as in the 1930s. Until the final days of her life, Begum Sarbuland continued to write devotional poetry about the Prophet and Medina.

The rise of modernist Islam was a global phenomenon that came to spectacular prominence in 1924–1925, when most of Arabia was conquered by the hardline Wahhabi-inspired Saudi leader Abd al-Aziz al-Saud. Under his rule, Mecca and Medina were almost completely purged of the types of devotional practices that formed the core of the Sarbulands' religious life. Graves and shrines were demolished, and a even practice as simple (and to the Sarbulands, as intuitive) as placing one's hand on the railing outside the Prophet's tomb were violently policed, and those who attempted it were pulled away or beaten. Begum Sarbuland's account is one of the few autobiographical narratives by an Indian woman in Arabia before the Saudi conquest.[23] In her preface, Begum Sarbuland fantasized of returning to Medina and clinging to the lattice around the tomb until all else faded away. Yet she knew well that this was no longer possible. Pilgrims were no longer able to worship in Mecca and Medina as she once had.

Muslim Women's Travel Writing in Colonial India

Travel writing emerged in India around the early to mid-nineteenth century, when it flourished with the arrival of the steamship, the train, and the printing press. By the 1880s travelogues were appearing everywhere, in books, in magazines, and in newspapers, where readers could find weekly updates on the peregrinations of their favorite writers and politicians. By the turn of the twentieth century, the travelogue had become a central pillar of Indian literary and intellectual culture and a key genre for negotiating, debating, and adjudicating the world and India's relationship to it.[24] This rich tradition offers bountiful sources for seeing the world through the eyes of Indians of the past.

Unfortunately, not all voices from that past are equally represented. Men overwhelmingly dominated the genre of travel writing in India, with women's writing representing only a fraction of the total number of accounts ever printed. One reason for this was that "among sharif [noble] women of the nineteenth century . . . knowing how to read was unusual enough, but knowing how to write was very rare indeed."[25] But then, becoming a published author took more than knowing how to write. Social restrictions meant that women who could write were expected to avoid full visibility in public spaces and thus often had nowhere to publish, at least under their own names.[26] Poorer and lower-status women like Amina bi were physically more present in public spaces but typically illiterate.[27] In short, the gender-segregated world of Urdu print offered little ground to women in the nineteenth and early twentieth centuries.[28]

The earliest Muslim women's travel writing was thus not printed but circulated in handwritten copies among family and friends.[29] Begum Sarbuland's diary is a case in point. She was initially unwilling to publish her travel narrative and seems to have been writing with an eye toward a familiar, intimate readership. By 1910, though, a few Indian Muslim women had begun to publish. As Sumaira Nawaz notes, as accounts increasingly began to appear in print, ultimately "the genres of roznāmchā [diary] and the safarnāmā [travelogue] became socially respectable and preferred means of self-expression" for women.[30] The earliest accounts were published by women from wealthy, liberal families. Most prominent were the women of the Tyabji family in Bombay—Atiya and Nazli Fyzee particularly—who left purdah and actively participated in India's public sphere.[31] There were also the queens of Bhopal—extremely wealthy and educated women who ruled their own state.

The first decades of the twentieth century saw other gradual but major changes that made it possible for everyday sharif (or noble) women to circulate their writings more widely. From the 1900s, Urdu-language magazines targeted at and increasingly produced by "sharif Indian ladies" began to pop up across the country.[32] A few women were even beginning to publish book-length accounts.[33] Despite having this option before her, Begum Sarbuland only allowed for her travel stories to circulate beyond her family orally, via speeches given only to sharif women in small, controlled settings. A 1914 magazine article makes a chance reference to one such gathering. The piece begins by noting that it had become an established practice for the "educated Indians" who visited Shimla annually to bring their wives and families with them. The author was "pleased to see" that women avidly attended all the happenings. The women did so in purdah but in such a way that they could "see the outside world" without being seen by it themselves. One such event was hosted by Begum Sarbuland: "I found the Indian ladies [who had come to Shimla] well informed on all current topics of the day, but the war [World War I] was the chief subject of conversation amongst them. I was particularly interested to hear Begum Sar Buland Jung of Hyderabad, at her own Pardah Party, give experiences of her visits to England, France, Germany, Russia, Turkey, Palestine, and Egypt, thus making the theatre of war quite vivid to us all."[34] Begum Sarbuland likely gave many talks like this after her return. She must have been an engaging raconteuse if her travel stories could still hold an audience four years after her return![35]

Other women of Begum Sarbuland's generation made similar choices about how and when to publish. One of these was Ummat ul-Ghani Nur un-Nisa (1885–1915), who like Begum Sarbuland was born in Hyderabad to an aristocratic family in the late nineteenth century. Coincidentally, the two women performed the hajj the same year, departing from Hyderabad just nine days apart. Then, they followed roughly the same idiosyncratic itinerary in the Middle East and even stayed in the same "Muslim hotel" in Damascus (which they both reviled). They may well have encountered one another in Mecca or Medina, where people of the same regional or linguistic background often met. Both women kept travel diaries, neither of which were published as books nor excerpted in popular women's magazines like *Tehzib un-Niswan*. Ummat ul-Ghani did, however, publish her diary in a handwritten family newspaper that circulated only among her relatives. In this way, her writing never left the bounds of her own family within her lifetime (though it was later recuperated from family records and published).[36]

At the opposite pole was another Hyderabadi contemporary, Sughra Humayun Mirza (1884–1958), who was eight years Begum Sarbuland's junior. Today Sughra is known as one of the founding figures of women's writing in Urdu. She published more than a dozen novels and at least five book-length travel accounts, the first appearing in 1914. Like Begum Sarbuland, she was a prominent supporter of women's education and social reform. However, while Begum Sarbuland preferred to work from within women's spaces and without publishing her views, Sughra took an outspoken approach and ultimately became well-known nationwide.

Outside of Hyderabad, Begum Inam Habibullah (1883–1975) was another exact contemporary. A sharif woman from northern India, she traveled to England in 1924 to visit her three young sons who were at boarding school there. Like Begum Sarbuland, she learned English and became closely involved in social reform. While in England, she too kept a daily record of her trip, partly for herself but also so that she could share impressions of the journey with her "sisters" back home. After returning to India, Begum Habibullah, who was still living in purdah, only shared the diary privately. However, as with Begum Sarbuland, her views clearly shifted with the passage of time. By the mid-1930s she had left purdah. It was perhaps this shift that made her also decide to have her diary published for a public readership; it was likely printed in 1937, not long after Begum Sarbuland's.[37] Begum Sarbuland's account may well have inspired her own effort.

Both women's decisions were by then not particularly radical. The 1930s was a time when more and more women were leaving purdah and the risks of appearing in public continued to decrease. (Khurshid Begum, the Sarbulands' eldest daughter, abandoned purdah at around the same time.) It may be that this more relaxed social atmosphere inspired women like them to share their experiences and impressions beyond their immediate circles.

Begum Sarbuland's Life

Begum Sarbuland spent her life surrounded by the rich, famous, and powerful. Outside the home, she headed a wide range of social organizations, promoted reformist causes, and met with caliphs and queens. She had easy, informal access to the ear of the nizam, the ruler of Hyderabad, and was tied to many of his most important officers and advisers by blood or relation. Her family included a range of famous men. These men were the subject of biographical tomes. They also wrote extensively themselves. Given her position within this matrix of power and print publishing, one might imagine that Begum Sarbuland would appear regularly in historical sources.

In fact, Begum Sarbuland only makes the most fleeting appearances in the printed historical archive. This is not because she was meek or chose to stay at home: to the contrary, she was a vocal reformist, an avid traveler, a woman who had no compunction about mounting a donkey for a ride into town. However, she had little interest in preserving her legacy; her contemporary family members rarely wrote about her either. Her birth, for instance, is not recorded in her father's autobiography. Her husband typically only mentioned her in passing in his own autobiographical works.

By contrast, the human archive she left behind was immense. Begum Sarbuland was survived by many children and grandchildren who loved and adored her. Much of what I learned about her, and particularly her later life, came from two years of interviews with her descendants, some of whom lived with her in the 1940s and 1950s. Family records include a huge trove of photographs, heirlooms, and even Begum Sarbuland's own handwritten diaries. The following section is based on scattered references to her in archival sources, on her own diary entries, and on interviews with her descendants. In this introduction, I only offer a broad sketch of her life to contextualize her travelogue. Readers wishing for a deeper engagement with Begum Sarbuland's life and work should read the biographical chapter, "Begum Sarbuland: A Life Untold," before reading the translation of her travelogue.

Akhtar un-Nisa (who would later bear the honorific title Begum Sarbuland Jung) was born in Hyderabad in southern India around 1876. However, her ancestral roots lay in Delhi. Her father was Nawab Agha Mirza Beg (1848–1933), a descendent of the Mughal royal family (plates 2 and 3). He was still a child when his family left Delhi in 1857 to flee a violent revolt against British colonial influence. Her mother, Sikandar Zamani (b. 1857), was also of royal blood (plate 4). The events of 1857 meant that both of Akhtar un-Nisa's parents (who were betrothed cousins) grew up in exile from their ancestral homes. Mirza Beg spent the rest of his childhood in Lucknow. After completing his schooling, which was conducted largely in English, he migrated to Hyderabad; Sikandar Zamani joined him later. In Hyderabad he rose quickly through the ranks, coming to wield enormous power as the tutor and later the trusted confidant of the state's ruler, the nizam. From childhood, Akhtar un-Nisa watched on as Hyderabad transformed from a Mughal successor state into a paragon of Muslim modernity, partly under her father's guidance.[38]

In 1894, Akhtar un-Nisa traveled to Delhi—possibly for the first time in her life—to marry Nawab Hamidullah Khan, a young lawyer who was famously the first student enrolled in the Muhammadan Anglo-Oriental College in Aligarh (plates 5 and 6).[39] This institution, which became Aligarh Muslim University, was meant to provide a modern colonial education to the hereditary landowning classes of northern India. From Aligarh, Hamidullah went abroad to study at Cambridge for several years before returning to India to practice law. Yet, for all his own personal successes, it was his marriage to the daughter of Hyderabad's most powerful official that unlocked a brilliant career in the state's judiciary. His new father-in-law helped him find promising employment and arranged for him to be given the honorific title "Sarbuland Jung" (*Sarbuland* literally means "head held high"; "*Jung*" or "Jang" was a marker of titles from the nizam.) From then on, Akhtar un-Nisa became known as simply Begum Sarbuland Jung (lit., "the wife of Sarbuland Jung").

The couple lived together in Hyderabad for several decades, producing at least thirteen children, of which ten survived (plates 8, 9, 10, 13, 15, 16, 18). In 1912, Nawab Sarbuland took an early retirement from the judiciary and moved with his wife to Allahabad (in the north of the country) and later to Delhi. He passed away in 1930. For the following two and a half decades, Begum Sarbuland lived variously in Delhi, Hyderabad, and Solan, sometimes alone and at other times with her children. Begum Sarbuland was in Delhi in 1947 when the partition of India and Pakistan gave rise to riots, killings, and mass displacement in the city. She not only witnessed this brutality firsthand but was herself displaced, though she found a safe berth with her son Halimullah, who temporarily rented an apartment for his family in Connaught Place. Her longtime home in the Daryaganj neighborhood of Old Delhi had been incorrectly declared "evacuee property" and was thus confiscated by the Indian government for allocation to a Hindu or Sikh refugee.[40] Its recovery took several years of legal battles. She passed away in 1957 and was buried in a plot beside her husband near the walls of Delhi's old city. Their graves are unmarked but can still be pointed out by the cemetery caretaker (plate 30).

From the splendor of nineteenth-century Hyderabad to the bloodshed of a sundered nation in 1947, from the royal palaces of the Hyderabadi nizam to a rented flat above a British-style pub in Connaught Place, Begum Sarbuland saw the highest of India's highs and the lowest of its lows. The span of her life was such that researching her biography led me through a wide swath of history. Conducting this work, I found myself variously examining dusty nineteenth-century sources and asking her surviving relatives via WhatsApp what it had been like to grow up with such a remarkable woman. Meeting with Begum Sarbuland's grandson in New Jersey and New Delhi to hear his memories of her, or listening to her granddaughter describe what it was like to witness the chaos of partition alongside her, the proximity to the past sometimes felt astonishing; it was as if I could hear a chain of voices that led directly back to the Mughal Empire itself. After all, Begum Sarbuland would have heard similar stories of chaos in Delhi from her own father, who had witnessed the toppling of the final Mughal monarch in 1857. In this way, Begum Sarbuland and her family seem to bind together two centuries of Indian history. Her life and writing reveal fascinating and important new details about India's historical ties to the wider world and how Muslim women in India responded to the broad political, social, and religious changes that were taking place in Islam.

Conclusion

The heart of this book is Begum Sarbuland's own words as they were preserved in her 1909–10 travel account. Further details of her trip and her life are found in the second part of the book, which contains an expansive biography, a translation of an advertisement for the book, and selected passages from Nawab Sarbuland's own account of the journey. If her travelogue provides a snapshot of her views in 1909–10, her biography reminds us that she was a dynamic individual who changed with the world around her. For instance, in the 1880s, she, like most girls of her time, was not taught English. Yet, by the time she began her travel diary, she had clearly mastered the language sufficiently to chat comfortably with British women, including the future British queen. Another example of her dynamism comes from her shifting veiling practices. In her diary, she muses often on what type of veiling best fit the various situations she found herself in, adapting her practice to the circumstances. Yet her calculations also changed over time. In 1909 she often chose to wear a burqa and a full face covering. By the 1930s she had transitioned to a full-sleeved sari blouse under a long coat, her head covered by a dupatta. Photographs of her ranging from 1910 to 1957 reflect these shifts.

There are several other differences emerge from reading her 1910 text alongside the 1935 preface. In her earlier writing, she was most interested in comparing the lives of Indian Muslim women with their counterparts in the Middle East. By 1935, though, she was more concerned with making comparisons to European women; the positive views she expressed about contemporary Syrian women's lives and independence in her diary fell to the wayside as she worried about what she called the rising tide of Westernization among Indian women. In her preface, she found new models to aspire to emulate in the Muslim past. She particularly points to Ayesha, the Prophet Muhammad's youngest wife, whom she describes as simultaneously devoted to her family and to assisting her society outside the home, as the ideal model for modern women. There is a photograph of Begum Sarbuland from the 1930s that seems to embody this attitude (plate 11). In it, Begum Sarbuland stands—fittingly enough—with her head held high, looking forward intently. She is dressed in a sari, with the tail end draped over the back of her head. In her hand, she holds a type of black cloth, probably a veil. Often in such pictures, the subject poses with a symbolic object—a stack of law books, in the case of Hamidullah. That Begum Sarbuland chose to pose with her veil but not wear it reflects her careful commitment to the institution of purdah without giving up her direct engagement with the world. This approach to purdah was partly shaped by her experiences in Europe and her reflections on the lives and practices of other women around the world.

Reading her 1910 diary against her 1935 preface, the photographic archive she left behind, and her biography offers a multilayered picture of Begum Sarbuland's personality and enduring dynamism. Rich with women characters, her diary also offers potential avenues to think more broadly about the lives of women in the society she inhabited. Most of the women that she mentions have left no record behind at all. These include her mother, Sikandar Zamani, her servant Amina bi, and her friend Mrs. Mumtaz. Begum Sarbuland's candid account of her interactions with each of them gives us a better sense of what their everyday lives were like and how they supported one another with friendship, advice, and practical assistance. For instance, Begum Sarbuland would probably never have made this journey if she were unable to leave her many young children at home with her mother, sisters, and servants.

The material history of her diary offers further insights. While Begum Sarbuland's voice was preserved through her published diary, it might have easily disappeared. Books in India quickly succumb to the region's fluctuating heat and dampness if not carefully cared for. Meanwhile, South Asian libraries have a poor track record of preserving women's writing and Urdu literature in general. Of the few hundred or so copies, only three are known to survive today. The first is in a private family archive in Boston maintained by Begum Sarbuland's grandson Abid Ilahi. The second copy belonged to the eminent historian of Muslim women in colonial India, Gail Minault. She acquired it at a secondhand book bazaar in Hyderabad and later donated it to the library of the University of Texas at Austin.[41] The third belonged to an Indian scholar, Maulana Nazir Ahmad. In 1982, his personal library was donated to Jamia Hamdard, a university in Delhi, where one copy of Begum Sarbuland's book remains today.[42] This is where I initially encountered it, lying next to a dead pigeon and covered with a film of dust.

In addition to her travelogue, the final piece of material history that Begum Sarbuland left behind in the family archives and that I include in this book is a rich collection of images. These images constitute an alternative, visual history of Begum Sarbuland's life. The photographs show

Begum Sarbuland between 1910 and 1957. Some are candid; others, formal. Each image tells a story of its own, whether through objects included in the frame or through her selected outfits and adornments. They reveal her shifting sartorial style, her attention to fashion, and her attitudes toward purdah. The collection also includes further images that contextualize her life, including photographs of her parents, husband, siblings, children, grandchildren, and family servants. Nearly all the photos underline the central role that family had in her life. The earliest show her with her own children, but it is her grandchildren who are most prominent in the later images. This gradual shift visually locates Begum Sarbuland in time and affirms that she cannot be read as a disembodied voice from the past. Her legacy continues to live on through her children and grandchildren. The oldest person pictured in this collection is her father, who was born in 1848. Many of the children who appear with her in the later photographs are alive today and living in India, Pakistan, Saudi Arabia, and the United States. This visual archive thus not only explores her past but also binds her story to the present.

Finally, I offer a brief note on my translation and the structure of this book. This travelogue was intended for a general audience, and I have attempted to preserve that accessibility by remaining faithful to her informed but informal tone. I have refrained from prefacing her account with extensive biographical and historical information that would be useful for researchers and scholars but distracting to lay readers. Those who wish to hear directly from Begum Sarbuland may turn the page and dive directly into her travel diary. Readers desiring a more detailed and nuanced picture of Begum Sarbuland's life should first read the biographical chapter following the translation, "Begum Sarbuland: A Life Untold." The chapter also includes brief biographies of Begum Sarbuland's father and husband, both of whom were profoundly influential figures in her life. The book concludes with a photo archive, a translated advertisement for the original travelogue, and selections from Hamidullah Khan's travel writing. There are few, if any, other instances of a husband and wife from colonial India both writing independently about the same journey. Comparing his style and views to hers sheds light on Begum Sarbuland's own account and shows how even closely related individuals might see and present the world differently. Finally, a glossary at the end of the book defines unfamiliar terms.

These scholarly additions to her account offer a path to understanding Begum Sarbuland's life and words from an academic perspective, but, ultimately, they cannot come close to matching the raw, emotive power of her prose. Even as Begum Sarbuland claimed that she was not herself a writer, she nevertheless asserted, correctly, that the "heartfelt emotions [in the book] will provide some guidance to my sisters." Quite beyond the historical value of this unique text, the heartfelt emotions of her travel diary retain the power to inspire and perhaps even guide her readers today.

Notes

1. H. Khan, *Safarnama-e Qustuntuniya*, 84.

2. Sumaira Nawaz argues that Begum Sarbuland's use of the Turkish burqa during her travels was a conscious mark of her mobility as a modern Muslim woman: "The Turkish burqa, much like contemporary calls for Islamic modernism, stood at the intersection of Islamic nostalgia and reform, illustrating the compatibility of the Islamic way of life with even the mechanised European steamship. The Turkish burqa was thus an outcome of the Muslim world's interaction with Europe's technological resources, which had no single outcome, no one sense of an 'omnipotent modernity,' nor a unified Islam." Nawaz, "East Is East," 18.

3. N. Akhtar, *Bibi's Room*, 43; Pernau, "Female Voices," 36.

4. Majchrowicz, *World in Words*.

5. Lambert-Hurley, *Elusive Lives*, 57.

6. Pernau, "Female Voices," 36.

7. A second benefit of avoiding the direct ship to Jeddah was that hajj ships from India were required to quarantine for seven days before landing. Thus, arriving in Jeddah from elsewhere in the region would spare them this discomfort. H. Khan, *Safarnama-e Madina*, 1.

8. H. Khan, 22.

9. This was a long-running theme in women's accounts of the hajj in Urdu. Quarantine regulations during the hajj pilgrimage required would-be visitors to go through a rigorous health check that included steaming one's clothes and washing in a public bath. Elite Indian women protested vigorously and even paid bribes to avoid being seen naked. These showdowns between Indian pilgrims and quarantine officials are a stock element of women's travel writing in the early to mid-twentieth century. For an engaging example, see a travel account written by Begum Sarbuland's sister-in-law, Rahil Begum Shervania. Majchrowicz, "Rahil Begum Shervaniya," 134.

10. Lambert-Hurley, *Elusive Lives*, 185.

11. For an overview of the history of travel writing by Muslim women, along with representative examples and translations, see Lambert-Hurley, Majchrowicz, and Sharma, *Three Centuries*.

12. While no pictures of Amina bi are known to exist, women who worked in similar roles appear in family photographs both candid and posed, underlining the enduring relationships that Begum Sarbuland had with those she employed (plates 8 and 27). Amina bi and her daughter reportedly remained close to the family until the end of her life.

13. For a similarly dismissive view from a prominent Indian theologian, see Shibli Nu'mani, *Safarnama*, 16–17. For an incisive examination of the racialization of Islam during this period, see Aydin, *Muslim World*.

14. On just one occasion, she refers to a woman she meets as an *araban mem*, or an "Arab mem." In this case, she seems to acknowledge the woman's Arabness but also others her by hyphenating her Arabness with an assumed Europeanness.

15. Green, "Spacetime," 408.

16. Green, 406.

17. Green, 405.

18. Green, 405.

19. For more on this important development in a global context, see Aydin, *Muslim World.*

20. On this point, see Majchrowicz, *World in Words*, 175–76.

21. On the conspicuous absence of these groups in Urdu hajj accounts from this period, see Majchrowicz, *World in Words*, 179–81.

22. On his life, work, and spiritual and political leadership, see Robinson, *Ulama of Farangi Mahall*, 151–76.

23. For an account of Indian responses to these events, and particularly to restrictions on devotional practices in Medina, see Majchrowicz, "Early Indian Responses."

24. For a detailed examination of the genre and its growth in Urdu particularly, see Majchrowicz, *World in Words.*

25. Minault, *Secluded Scholars*, 24. On this point, see also Naim, *Texts and Contexts*, 202.

26. On the topic of women publishing under other names than their own, see Minault, "Zay Khay Sheen."

27. For an assessment of the history of working-class Muslim women traveling to Europe, see Fisher, *Counterflows*, 222–24.

28. These restrictions were present, to a greater or lesser extent, in most Muslim societies around the world at the time. For a global overview of Muslim women's travel writing during the nineteenth and twentieth centuries, as well as a description of further causes for the limited availability of narratives by Muslim women, see Lambert-Hurley, Majchrowicz, and Sharma, *Three Centuries*, 6–27.

29. Majchrowicz, "Malika Begum's Mehfil," 870.

30. Nawaz, "East Is East," 6.

31. For an excellent review of the earliest Muslim women travel authors in India, see Fyzee-Rahamin, *Atiya's Journeys*, 3–6. See also the translated text in the same volume for a comparison with a publicly published account from 1906.

32. Minault, *Secluded Scholars*, 105–57; Alam, *Familial Intimacy*, 22.

33. Among them, the authors Atiya Fyzee and Sultan Jehan Begum. See Fyzee-Rahamin, *Atiya's Journeys*; Lambert-Hurley, "Out of India," 263–76.

34. Datta, "Comradeship," 289. The references to Germany and Russia are surely mistaken. There is no record or family memory of Begum Sarbuland having traveled to these countries.

35. On the prominence of oral travel accounts in colonial India, see Majchrowicz, *World in Words*, 64–65.

36. Ummat ul-Ghani, *Safarnama*. For a partial translation, see Lambert-Hurley, Majchrowicz, and Sharma, *Three Centuries*, 107–16.

37. Begum Inam Habibullah, *Ta'ssurat-e Safar-e Yurap.*

38. On various aspects of this transformation, see Beverley, *Hyderabad*; Datla, *Secular Islam.*

39. On the early life of this institution, see Lelyveld, *First Generation.*

40. Though the state eventually returned ownership to the family after years of legal wrangling, the case dragged on until after Begum Sarbuland passed away. On the lived history of evacuee property confiscation, see Zamindar, *Long Partition.*

41. Personal communication, April 5, 2023.

42. Today a digital version of the book made from this copy is available online at https://rekhta.org/.

PART 1

The World through a Woman's Eyes

A Travelogue of East and West

Begum Nawab Sarbuland Jung Bahadur

دنیا عورت کی نظر میں

مشرق و مغرب کا سفرنامہ

بیگم نواب سربلند جنگ بہادر

The World through a Woman's Eyes

A Travelogue Containing an Eyewitness Account of Mecca, Medina, Egypt, the Levant, Iraq, Italy, France, England, and Other Regions of the World

by

Begum Sahiba Nawab Sarbuland Jung Bahadur (Delhi)

Published by
Khwaja Book Depot
Khwaja Barqi Press
Jami Masjid
Delhi

PREFACE

WHAT IS LIFE? IT IS A JOURNEY. We have long been traveling on this journey, we continue on it now, and we will remain on it until we reach our destination, though there is no knowing if we will manage to reach our True destination.

What is the story of our life? It is a travelogue, a travelogue we have long been writing through our actions, which we are writing now, and which we will continue to write until our life's journey concludes. When it does, our entire life will appear as though it were nothing more than a dream or a story.

The present travelogue contains one interesting chapter from the story of my life. I was young then, and my husband, my lord, Nawab Sarbuland Jung, chief justice of Hyderabad (Deccan), was still alive. This travelogue relates a period when my life was still pervaded by the intoxication we call "life." Now, as I put these scattered pages into order, Nawab Sarbuland Jung, by whom I traveled the East and the West, has left this world. He is in the spiritual plane while I drown in the inundations of this material world. Though I am still surrounded by life's physical luxuries, my soul's tranquility departed with him.

I am deadened now to the pleasures of life. Even still, my soul grows restless when I recall the lattice that surrounds the tomb of the Ruler of Medina [the Prophet Muhammad], when visions of its green dome appear before my eyes. My heart yearns to travel to the Land of the Beloved, to return to the soil of Medina and press it to my eyes.[1] Just my lord's tomb, its lattice, and me. I would grasp the lattice and cry out, "My lord! My master! Why did you leave me in India to suffer without you? Oh, Father, are you angry with me? Yes, I am a sinner, but I belong to you. I am base, but I am yours. Summon me. Summon me to you. Living far away in India leaves me distraught." I would speak these words and shed these tears, and then I would hear his voice intone, "I am yours, and you are mine." I would hear these words and fall unconscious, unconscious forever more.[2]

Only the memory of Medina soothes my heart now. This travelogue began from my pilgrimage to Medina, and the journey of my life will conclude with my recollections of the Lord of Medina. These are the provisions of my life, and this is my travelogue. Before I had traveled, and before I had made a pilgrimage to the Land of the Beloved, I had longed to see it. And now that I have made not just one but two pilgrimages there, my separation from it fills me with sorrow. The pull of the Prophet's land drew me from India toward itself. Then, by the grace of the Prophet, I also saw Egypt, the Levant, and Iraq.[3] I visited Italy, France, and England. As I traveled, I went on recording my emotions and feelings.

In my travelogue you will find neither literary merit nor poetic flights of fancy. I am an emotional human; here you will find nothing but emotions. I have simply recorded my emotions in broken, ill-fashioned words. I myself do not know what I have written. The truth is that I did not write this travelogue with the intention of publishing it. This is not really a travelogue, so much as a private, personal diary I kept to preserve my memory or to assuage my emotions.

Children love their parents and everything about them, so it is no wonder that my children love this diary of mine too.[4] It is through their insistence and effort that my travelogue, or diary, has been published. In my opinion it is too lacking in literary finesse and style to be put before the nation.

Alongside my heartfelt emotions, in this travelogue you will also find sketches of the cultures and societies of women from other nations. From them you will learn to what extent the women and men of other countries differ from the people of India and just how distant their societies are from ours. There were many things in these foreign countries that I did not like at all. I also saw much that should be adopted by Indians, and which Indian bibis in particular should adopt.

Unfortunately for India, Western culture is spreading here. The women of India are charging madly toward Western freedoms. The thing I felt most strongly after traveling abroad was that the same blind freedom that is now destroying India has already plundered the peace and true repose of the West. This blind freedom and this foreign culture have made the visible aspects of their lives appear very attractive and pleasant. Yet inwardly their lives are not so pleasant at all but seem rather to be devoid of any spiritual contentment.

After closely examining the situation in India and other foreign lands, I have concluded that the best life for women is the one once lived by the Muslim bibis of the earliest centuries [of Islam]. The bibis of the first Islamic centuries did not remain shut up in their houses chewing *paan* and frittering away their time in useless chatter. Nor were they objects for men's entertainment, like Western women. They were responsible members of the household. They raised their children and maintained their homes. There were also women among them who, like Hazrat Ayesha, even went out to the battlefield to tend to the wounded.[5] There was no excess then, as today. This is why the generations that they raised in their laps soon conquered half of the world.

My message to my sisters is this: After reading my travelogue, please consider the conditions of women in these many countries and reflect on your chosen direction on the battleground of life.[6] I do not claim that my travelogue contains a vast wealth of new information, but I do believe that my heartfelt emotions will provide some guidance to my sisters.

The Dweller of the Dust of Medina,
Begum Nawab Sarbuland Jung
Daryaganj, Delhi

Notes

1. A gesture of veneration and humility.
2. As Begum Sarbuland surely knew from her second trip to Medina in 1934, her vision was no long possible in reality. Under Saudi rule, such forms of devotion as clinging to the lattice surrounding the Prophet's tomb were strictly prohibited and heavily policed. Saudi regulations, though, had no effect on her written expressions of devotion. See Majchrowicz, "Early Indian Responses."
3. The diary mentions Iraq twice, including on the title page, but she never visits any location that could conceivably be considered part of Iraq. The closest she reaches is Damascus.
4. This line suggests that the diary was read by and known to her family, not a "private diary" in the restricted sense of the phrase.

5. Ayesha (614–678), or Aisha, was the Prophet's third wife. She is acclaimed among Sunni Muslims particularly as a scholar, intellectual, and leading figure in the spread of Islam.

6. This sentence suggests, perhaps intentionally, that women were Begum Sarbuland's intended readership. This was a common assertion in travel writing by women authors in Urdu. Language of this kind may have helped address questions of propriety surrounding publication for a general readership, but it may also be that Begum Sarbuland wished to speak primarily to sharif women like herself. Majchrowicz, "Malika Begum's Mehfil," 871.

CHAPTER 1

HYDERABAD

حیدرآباد

The Pull of the Prophet's Land

I used to wish that I could grow wings and fly to the green dome [of the Prophet Muhammad's tomb] so that I could press my eyes to the blessed soil of the Land of the Beloved [the Prophet] (fig 4.1). This wish was little more than a yearning in my heart until one evening in 1909 (1327h, 1319f).[1] I had finished my housework and gotten into bed when Nawab Sarbuland Jung sahib suddenly said to me, "Would you like to go on hajj this year?"

Me (exuberantly): Inshallah, I will absolutely go!

Nawab sahib: Where will you leave the children, and with whom?

Me: I will leave them with my dear mother.

That is what I said to him, but in truth I suspected that my mother would want to come on hajj with me. When I mentioned the matter to her the following morning, she generously and lovingly agreed to stay with the children. I was sad that my mother would not be coming, but I made my peace with it for the sake of the children. Then, and as God commands, we finalized our intention to perform the hajj and the pilgrimage [to Mecca and Medina].[2] We began our preparations in the month of Jamada II 1327h [June 1909].

Meanwhile, our dear friends offered up various reasons to try to dissuade us from going. Even my father, Muhammad Agha Mirza Beg Sarvar ul-Mulk, [the former] tutor to Nizam Rukn [Mir Mahbub Ali Khan], forbade us, saying, "Go next year. Postpone your plans for now."[3] But the Beloved's tomb had already begun to draw us to it, and despite their determination to stop us, we could not be stopped. We remained undeterred. Soon our luggage was packed, and Nawab Sarbuland Jung's request for work leave was magnanimously approved by His Highness [the nizam].

Gatherings with the Nawab's Friends

Nawab Sarbuland Jung Bahadur began to host his friends and lawyers from the high court for meals from the first of Ramadan 1327 [September 16, 1909]. Some he invited to dinner; others he hosted at "breakfast"; for others, he threw "parties." Guests were invited one or two hundred at a time. These gatherings continued unabated until the fifteenth of Shawwal.[4]

Our Farewell Gatherings

From the sixteenth of Shawwal [October 31], we began to attend farewell gatherings hosted by our friends and family.[5] I will put the men's "dinners" and "breakfasts" aside and mention only the women's gatherings held in my honor. The first dinner and breakfast included my entire family, male and female. It was hosted by Colonel Nawab Sir Afsar ul-Mulk's wife and the colonel himself.[6] I was hosted inside the house with my sisters and sisters-in-law. Nawab Sarbuland Jung was in the outer rooms with my brothers and brothers-in-law. There were many other guests too, male and female. Begum sahiba, whom I call Khala Jan, asked me to come at nine in the morning. She had me stay until evening. She tied an *imam zamin* to me before I left.[7]

Meanwhile, Colonel sahib kindly took Nawab sahib to the palace in his motorcar so that he could present himself there for a salaam and to give a departing offering. He remained there the entire night, until nine in the morning. In the morning he had chai on the blessed veranda with Colonel sahib. He stayed the night [at the colonel's house]. In the morning, Nawab sahib had tea on the palace veranda with Colonel sahib. At nine, he presented his parting gift to the sovereign, then went straight to the high court and came home again at the usual time.[8]

A Visit from Lady Hydari

17 Shawwal [November 1]: Mrs. Hydari visited today. She is a dear sister of mine.[9] She has been visiting me regularly since I decided to travel. She is very sad at our impending separation. She tried to dissuade me, saying,

"Don't go now, go another time." Since this journey was willed by God, her words had no effect on me. Our resolve held firm.

Some Hyderabadi Officials Revolt

Around the time of our departure, there was a revolt among the state's officials. The unfortunate Aziz Mirza was removed from his position and ordered to leave the city.[10] Our friends and family were certain that His Highness would require Nawab Sarbuland Jung to delay his departure, but God's decision was not set aside. His Highness happily accepted his [departure] offering and allowed him to leave for this blessed journey.

At this time, all my children [were in Hyderabad] except for Mahmudullah Khan, who was nine years old and had already been in London with Khwaja Abd ul-Majid for about six months (plate 7).[11] Khwaja sahib is Nawab sahib's maternal cousin. He is studying to become a barrister in London (fig. 1.1).[12] My eldest son, Muhammadullah Khan (plate 10), and Masudullah Khan (plate 8), my middle son, are both home on holiday from Panchgani, where they attend school.[13]

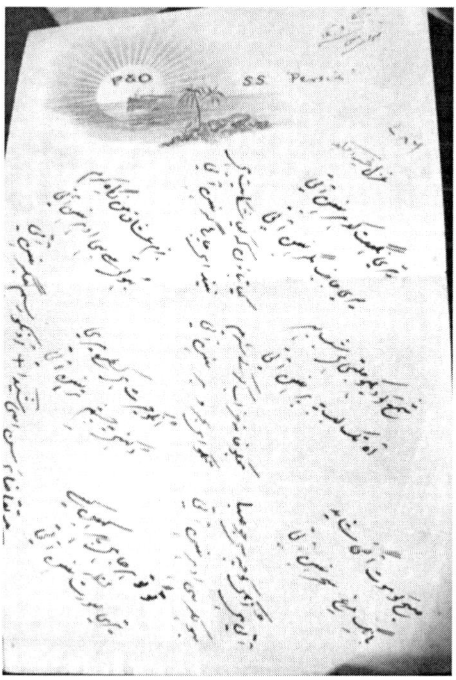

Fig. 1.1 In May 1909, Abdul Majeed Khwaja set sail for England, reportedly hoping to make a name for himself and earn the hand of Begum Sarbuland's eldest daughter, Khurshid Begum. While at sea, he composed poetry for his future spouse on stationery from the ship and mailed it to her in Hyderabad. The poem, a ghazal, is signed with his penname, "Shaida Khwaja." *Source: Shakira Sultana.*

My Girls

My eldest daughter is Khurshid Begum (plates 8 and 13).[14] She is thirteen years old. She is studying in an excellent zenana school that I and several other educated ladies had started with great effort.[15] These days Janab Mr. Casson Walker is working as the [state's] financial adviser. His memsahiba, Lady Casson Walker, is the school's president.[16] Lady sahiba is capable, generous, and affectionate. I call her "my dear mother." She refers to me as "my dearest daughter."[17] Her two daughters, Dessy and Sheila, are here now. The older daughter, Dessy, was married six months ago and now goes by the name Mrs. McClintock. She will be moving to Burma in three months. I am upset that she will leave Hyderabad before I return. Lady Casson Walker sahiba has been a major help to the zenana school. She alone looks after its organization and operations. Khurshid Begum was the first pupil whom Lady Walker kindly enrolled in the school.

I wish that I could take Khurshid Begum on the trip with me, but Nawab sahib said it was best that she not come. It might also negatively impact her studies. Accordingly, I have left Khurshid Begum here with my other children.[18] My mother has come to stay in our house with them. My younger sister Aziza Maryam Sultan Begum and her spouse, Muhammad Ahsan Khan, have also moved to my house for the time being. She will give birth soon. For the past two years, I have had her stay at my house when her due date was near so that she would not be alone. She lives near my house, Iqbal Manzil. Iqbal Manzil is rented; I do not own it. I am happy and relieved that my sister, sister-in-law, and her little girl will be staying there.

My Sister-in-Law's Sorrow at My Departure

Nawab sahib and I are ready to depart. My middle sister-in-law, Sikandar Dulhan, is distraught at the thought of our separation. She loves me very much, and I love her too. She often comes to stay with me. Her husband, my middle younger brother, Muhammad Mirza Sajjad Beg Nawab Usman Navaz Jung Bahadur, is studying medicine in London Adabza.[19]

The Prime Minister's Affection[20]

19 Shawwal [November 3], *Tuesday*: His excellency the prime minister invited Nawab sahib over and gave him an imam zamin of five rupees and two ashrafis. He affixed it himself with great kindness before seeing him off.

20 Shawwal [November 4]: I was invited to "dinner" at the prime minister's home. Ghausiya Begum and Rani sahiba, the maharaja's senior queen, kindly hosted me.[21] It was a wonderful dinner. The food was rich and delicious. There were some other guests too. When it was time to leave Ghausiya Begum, who is called Musarrat Mahal, and Rani sahiba, called Rahat Mahal, each gave me an imam zamin of one ashrafi and four rupees. I thanked them and returned home at 1:00 a.m.

Farewell Meetings

21 Shawwal [November 5]: I stayed home today. Three bibis from my pir's family came to visit. These bibis each gave me an imam zamin, had food, and left.

22 Shawwal [November 6]: My dear friend Mrs. Karim Khan invited me to "breakfast." I went alone. There were just one or two bibis from her family present. The breakfast was very good. She gave me an imam zamin when I left. I thanked her profusely. I left at 2:00 p.m.

Shrine Visits

After the *maghrib* prayer, my mother [Sikandar Zamani Begum], Khurshid Begum, and I went to the shrines of Hazrat Yusuf and Hazrat Sharif to recite a fatiha and request leave to depart.[22] Whenever I visit there, I always take the route that runs behind the shrines. I had the carriage take this route as usual, but this time there was a large pile of wood blocking the road. Using the lanterns I had brought with me, the three of us struggled mightily to climb over the wood and reach the shrine. We read a fatiha and stayed a short while before going to the mosque. We prayed two *rakats* upon entering.[23] Then we went to the shrine of Shah Khamosh, which is nearby.[24] We read a fatiha, prayed *isha*, and returned home.

23 Shawwal [November 7]: Today I decided to visit the shrine of Baba Sharaf ud-Din and take his permission to depart.[25] In the morning I wrote to my dear friend Mrs. Mumtaz, "Sister, today I want to go to Baba sahib's hill to recite a departure fatiha, please join me. When you come, please ask for permission to take out Khalu Jan's motorcar." I then wrote to Mrs. Karim, saying, "You please also come along and join me at Baba sahib's hill," because she had been wanting to see [the shrine]. I also sent a letter to my dear sister Amina Khatun, but, unfortunately, she could not come because she had some task to attend to that day. My dear friends Mrs. Karim and Mrs. Mumtaz came in the motorcar around eleven. My middle sister-in-law, Iskandar Sultan, and my mother joined the party. We were too many people, and there was only one car, so we sent Mrs. Karim, Iskandar Sultan, and Mother ahead first. The rest of us went after.

Lunch at the shrine had already been arranged. Thus, we arrived there in waves and then had lunch. Mother, Mrs. Mumtaz, Mrs. Karim Khan, Iskandar Sultan, Khurshid Begum, and I all climbed the hill to the shrine barefooted. Mrs. Karim Khan was exhausted, but we all reached the sacred place and prayed a fatiha. Everyone prayed that Nawab sahib and I would travel and return home safely. Then we prayed two rakats in the mosque, followed by the *zuhr* and *asr* prayers. We distributed coins and sweets to the fakirs and began the walk down. Mrs. Karim was in a rush to go back, so she left first with Mother and Iskandar Sultan. Mrs. Muntaz and I stayed until the maghrib prayer. When the car came back for us, we returned safely home. I gave my heartfelt thanks to Mrs. Mumtaz and left. Mrs. Karim Khan had already been taken home.

A Gathering at My Elder Brother's

24 Shawwal [November 8]: My elder brother and my sister-in-law invited my children and me to breakfast.[26] All my brothers, sisters, sisters-in-law, and brothers-in-law were there. The food was excellent, delicious. After breakfast my brother and Nawab Sarbuland Jung went to the high court. We stayed until teatime. We left after having tea. We were given imam zamins.

25 Shawwal [November 9]: My elder brother Zulqadar Jung Bahadur also threw a separate departure party for Nawab Sarbuland Jung.[27] We both thanked him.

Dinner at Lady Hydari's

26 Shawwal [November 10]: My dear sister Lady Hydari Amina Khatun invited Nawab sahib and me to dinner. Only Nawab sahib, Khurshid, and I were at this dinner, because I have begun to appear before Mr. Hydari [unveiled]. Since he will be present at the table, none of my sisters or sisters-in-law were invited. Mother was also unable to come. Mr. Hydari only invited those bibis who do not wear a veil before Nawab sahib. Only those bibis were present who could appear before both Mr. Hydari and Nawab sahib unveiled.

In any case, we sat at the dinner table, which was beautifully decorated with silver flower vases, each at our proper places. It was a huge meal that lasted until midnight. Allah, Allah! The dinner finished, and we all stood up and retired to the drawing room to talk. They gave us imam zamins. We thanked them profusely and left.

The Day of My Departure

27 Shawwal [November 11]: It is the day of my departure. Khurshid is crying a lot. I try to console her, but she is very sad. My mother is also very sorrowful at our impending separation. I consoled her too. On this very day I mailed a letter to my father in Lucknow.[28] I wrote, "Please give this base one permission to depart. Inshallah, if I do not lose my life, I will quickly return to kiss your feet. Please do not worry." I also sent a letter to my pir Maulvi Muhammad Qiyam ud-Din Abd ul-Bari.[29] This great personage, the younger son of my pir Maulvi Muhammad Abd ul-Wahhab, is a true pir. No one alive today is his equal. I requested his permission to perform the hajj too.

Once I had finished all the necessary arrangements for my trip, all my friends and family came to see me off. By 2:00 p.m., everyone was there. It was a striking scene. Everyone was afflicted by my separation from them. Some had even come a day early. Mrs. McClintock also brought an imam zamin. Mrs. Walker brought me some excellent cow ghee for the journey. Mrs. O'Dwyer, the Resident's bibi, also came.[30] Mrs. [Henry Daly] Griffin and Mrs. Hogalf both kindly came to see me off. These memsahibs are my friends. I forgot to write that each of them delighted me by having me over for tea or dinner. Mrs. Griffin is the bibi of an Allahabad High Court judge. She is visiting Hyderabad for a few days. Mrs. Hogalf is the bibi of an army officer. They left around four; I thanked them all profusely for their kindness.

The moment of my departure was quickly approaching. My heart was exceedingly agitated, but I remained outwardly calm. My middle sisters-in-law, Mrs. Haidar Jivan Beg and Mrs. Usman Nawaz Jung, placed some beautiful flower garlands around Nawab sahib's and my neck.[31] Mrs. Hydari and the others did the same. Outside, Nawab sahib was given several more garlands.

Then it was time. I went downstairs and got into the motorcar at seven. All the servants were saddened by my departure. I embraced my mother and said goodbye to my sisters and sisters-in-law. I hugged Khurshid and gave her my affection. Then I said goodbye to everyone and sat in the motorcar. Today, Muhammadullah and Masudullah will go back to their school in Panchgani. They will travel with me until Bombay. Masudullah sat with me; Muhammadullah left with his father.

Suddenly, before we could pull away, my friend Mrs. Mumtaz ud-Daula appeared. I leaped out of the motorcar to meet her. She lovingly tied an imam zamin on my arm. She had her daughters La'iq Bibi and Khaliq Bibi with her. Bano Begum, the teacher at the Girls' School, also came to see me off. I thanked my sister Mrs. Mumtaz profusely, said goodbye to everyone, and departed. My neck was laden with floral garlands.

So many people had come to Hyderabad Station to see Nawab sahib off that it would be very difficult for me to depart from there. There were nearly five hundred people present. That is why I went to Begumpet Station to board the train.[32] My dearest sister Mrs. Hydari and Mr. Hydari came. I spoke to them for about half an hour.

The Train Arrives

The train arrived. I said goodbye and climbed aboard. Coming with Nawab sahib from Hyderabad Station were my brothers Mirza Haidar Jivan Beg and Mirza Akbar Beg Khan,[33] my brother-in-law and cousin Nawab Nasrullah Khan, my younger brother-in-law Muhammad Ahsan Khan, and my cousin and brother-in-law Mahbub Sultan. They all traveled as far as Begumpet Station. The number of garlands around Nawab sahib's neck was beyond reckoning. Everyone had given him one. The train began to move. I again shook hands with Mr. and Mrs. Hydari and my brothers. Akbar Beg was very hungry, so I quickly passed him some samosas from the tiffin basket through the window of the moving train.

Our compartment is a first-class family carriage decked with every necessary convenience. Right now it is 9:00 p.m. Muhammadullah and Masudullah are with me, as are their servants, Fayyaz Ali and Sheikh Husain. The carriage is very comfortable. It has a vestibule blocked by a hanging green screen, as is appropriate for a carriage with women travelers. There is a car for the servants attached to ours. The train is very comfortable. Once we began to move, we opened the tiffin basket, laid out food on the table, and ate. Then we did *vuzu* and prayed isha. I easily put both children to bed. Nawab sahib rested after saying his prayers, and I fell asleep too.

Gulbarga Station

28 Shawwal [November 12]: The train arrived at Gulbarga Station at 4:00 a.m. We had our carriage detached from the train because we wanted to go to the shrine of Khwaja Banda Navaz.[34] I prayed the morning prayer, then had breakfast and wrote two pressing letters to Hyderabad. Nawab sahib also wrote two important letters. By then it was eight. Two carriages belonging to the *subedar* of Gulbarga, a landau and a brougham, came to collect us.[35] We were escorted by four cavalry and a few assistants. Muhammad Salim Beg, the grandson of my aunt Anjuman un-Nisa Begum, got ready to take me to the shrine of Hazrat Khwaja Banda Navaz. Masudullah, Salim Beg, and I sat in the brougham. Nawab sahib and Muhammadullah sat in the landau. We set off, accompanied by the cavalry and the assistants. We arrived at the shrine after passing through the beautiful little settlement near the shrine.

Visiting the Shrine

We got down from the carriages and presented ourselves at the shine with bare feet. We prayed a fatiha. Then we prayed fatihas at several other shrines. We prayed two farewell cycles at the mosque. After completing this pilgrimage, we prayed the Friday prayer before leaving to see a fort. I also saw a mosque I liked. It is a beautiful building, but, sadly, both the fort and the mosque are in the wilderness and far from town.[36] We toured the fort and mosque for a full hour. The mosque was very clean. We prayed two *nafl* rakats here too.

Return to the Station

We left for the station again around 2:00 p.m. Everyone was very hungry. We ate lunch with Salim Beg. The food was sent by the subedar. He also brought an imam zamin and garlands himself. My *sarkar* accepted all of the imam zamins that his friends wanted to tie to him in the men's section and then returned them again with thanks. But no one would take them back from me. I alone was given imam zamins worth six hundred rupees. Allah! Allah! What wonderful friends God has given me. They love me deeply and were profoundly affected by our separation. Although I was intoxicated by my desire to see the Holy Places, I was profoundly affected too.

The stationmaster quarreled with us when we arrived at the station. He refused to attach our carriage to the mail train, and I had no choice but to give him an offering of four rupees, which I passed to him via Salim Beg. I didn't tell Nawab sahib about this; if he knew, he would have surely stopped me. Anyhow, after this the stationmaster attached our carriage to the mail train, and we departed Gulbarga.

Arrival in Bombay

29 Shawwal / November 14, 1909;[37] Our train arrived at Bombay station at 5:00 a.m. We were met by Nawab Sarbuland Jung's nephew Mahmud

Ahmad. I've known him since childhood. He often stayed with me. I said, "Hey, Mahmud Ahmad, where have you been?" He replied, "I'm going on hajj to the house of God too!" I replied, "Great, let's go. Is there anything better in this world than that?"[38]

Mahmud Ahmad hired a carriage and picked up my bag himself. With Nawab sahib's permission, the two of us got down from the train and went straight to the carriage. Muhammadullah and Masudullah came with us. Nawab sahib gave the order, "You all go straight to the Taj Mahal Hotel, and I will come afterward."[39] We set off and arrived at the hotel. It was packed. The manager said, "Janab, there are seven hundred guests here right now. We don't have any empty rooms. I'm so sorry."

We then went to the Apollo Hotel, which is nearby, but there was only one room available. We took it anyhow. The owner charged us a lot; we ended up paying eight rupees per person. I ordered some food and went to the room. Soon Professor Muin ud-Din Ahmad came to call. He is one of Nawab sahib's closest friends and a very accomplished man. I had a chat with him. During our conversation the subject of photography came up. I said, "Professor, kindly buy me a Kodak camera that I can easily carry with me on the trip." I also requested some other things I needed. He said, "I'll bring them now," and went straight to the bazaar.

After he left, I took a bath and changed. By then the table was set, so I went to eat with Masudullah, Muhammadullah, and Mahmud Ahmad. The food was good, and we ate well. Nawab sahib hadn't returned yet. He was busy arranging tickets for the ship and so had his lunch in a restaurant.[40] After lunch I sat to write some letters, including to my father and Maulana Maulvi Qiyam ud-Din Abd ul-Bari in Lucknow. I wrote another to my mother in Hyderabad. I also wrote to Khurshid Begum; Sikandar Sultan; my middle sister-in-law, Mrs. Usman Nawaz Jung; and my middle sister, Qarshiya Sultan.

I had only just closed my pen box when Nawab sahib came in and announced, "Well! Our tickets are all arranged. Move quickly—there isn't much time." I was soon ready, but the professor still hadn't returned. I took my handbag and went to the port with Mahmud Ahmad, Mahmudullah, and Masudullah.[41] Nawab sahib arrived later with our luggage. I went to wait in the waiting room. I began to think that when the professor arrived, I would need to take from him the things that I had asked him to purchase. But the waiting room was filled with men and women. I was wearing a very beautiful, moss-green Turkish burqa made of pure silk. I was still thinking about the professor when I suddenly saw him pass in front of me carrying the objects that I had requested he buy. Mahmudullah and Masudullah went over to call him.

"Wah janab! [Well, sir!]" he said. "There was so little time. I couldn't bring some of the items you requested. I got a few of them, though."

"I thank you for your kindness. I've given you a lot of trouble. Don't worry about the remaining items. Mahmud Ahmad will be coming to Mecca to meet me later; please send those things with him. I'll take these items now. Now I have the two blankets that I really needed." Then I thanked him again and said, "Please send Muhammadullah and Masudullah to Panchgani School today."

Professor sahib happily agreed to my request. Taking both my children with him, he shook my hand and departed. Then I went with Nawab sahib to another room, where the passengers are given a medical examination. I was only here for a short while; I showed the doctors my pulse and then left again. When I emerged, who did I see but Muhammadullah and Mahmudullah standing there with the professor. I gave both children my love and affection once more. Then I said as-salamu alaikum to professor sahib and left.

My First Steps at Sea

Then I got into a small boat. The strange thing is that I did not recognize that this was a boat because I had never seen a boat or a ship before. When I got into the boat, I mistakenly thought that this was another place we would need to wait in. Nawab sahib and I sat down on a bench. There were also many angrez and mems, some of whom were seated, while others were standing. My servant Amina bi sat down in front of me, by the luggage. I have only brought one servant with me. We left one wet nurse, Husain bi, and one servant, Khuda Bakhsh, and Mahmud Ali and Mahmud Ahmad behind, along with a lot of baggage. This is because we are going on the mail boat. These people, Mahmud Ahmad, etc. will come [later] on the hajj ship.[42]

Then the boat began to move. Nawab sahib had not told me that this was a boat. When it moved, I got frightened and said, "What is this place? Why is it rocking like this?" Nawab sahib replied, "This is a boat; it will take all of us to the ship." Then I started to take a closer look at my surroundings. The boat moved forward slowly. We traveled on until we reached the ship. The boat stopped. A ladder was brought from the ship and lowered onto the boat. We and the rest of the passengers boarded the ship.

Notes

1. Begum Sarbuland here gives the date according to the three calendars in primary use in Hyderabad at the time: the Gregorian, the *hijri*, and the *fasli*.

2. "As God commands" refers to the religious requirement to perform the hajj. There is no such requirement to visit Medina.

3. This exchange would have happened by mail, for her father was then in Lucknow.

4. The following month—thus, approximately forty-five days.

5. Her final son, Halimullah, was born on October 27. The failure to note this detail is difficult to explain.

6. Nawab Sir Muhammad Ali Beg, KCIE, MVO (1852–1930).

7. An imam zamin is a band with a token amount of money stitched inside. A sort of amulet, it is meant to protect the recipient during travel.

8. Mahbub Ali Khan, Asaf Jah VI (1866–1911) was then the nizam of Hyderabad. Begum Sarbuland's father, Agha Mirza, had tutored and advised him from his youth.

9. Amina Hydari (1878–1939). She was a member of the Tyabji clan and wife of Akbar Hydari (b. 1869), who began working in Hyderabad in 1906 and who would later become prime minister of the state. Amina Hydari helped found the Mahbubiya School where Begum Sarbuland's daughter was then enrolled. The use of the term "sister" is figurative.

10. Muhammad Aziz Mirza (1865–1912), a judge and official in Hyderabad. For a full but sympathetic account of his exile, see Muhammad Akbar Ali Beg, *Muhammad Aziz Mirza*, 53.

11. He would later go by the name Abdul Majeed Khwaja.

12. He traveled to England in May 1909. Family lore explains this journey's beginnings as follows: "Begum Sarbuland Jung's great granddaughter Farida Akbar relates that her grandfather Abdul Majeed Khwaja as a teenager was horse riding in a forest when he beheld Begum Sarbuland Jung's eldest daughter Khurshid riding in the distance. Abdul Majeed was immediately smitten and requested an audience with Begum Sarbuland Jung and for her daughter's hand in marriage. Begum Sarbuland Jung told the young man that the fact he was an heir to an immense landed fortune gave her pause. She asked the young man to first obtain a professional degree so he would be prepared for the modern world, for example a bar-at-law from London. In 1909 Abdul Majeed set sail on the P&O SS Persia to study at Cambridge University." Fig. 1.1 shows a poem he wrote and sent to Khurshid while sailing for England. Abid Ilahi, personal communication, April 10, 2024, and Shakira Sultana, October 28, 2023.

13. Panchgani is a hill station in the mountains of contemporary Maharashtra, near Pune, about five hundred miles west of Hyderabad. It was home to several elite boarding schools.

14. This section heading refers to "girls," but only her eldest daughter is mentioned here.

15. Mahbubiya Girls' School. See the biography for more information. See also Minault, *Secluded Scholars*, 205–6.

16. Sir George Casson Walker (1854–1925) and his wife, Fanny Coates. From their arrival in Hyderabad, Coates had taken "an active interest in establishing contacts with the ladies of the nobility, and even presided over a society 'for promoting healthy social intercourse and improvement.'" In 1906 this group requested funds to start a school for elite Muslim girls. This effort culminated in the establishment of Mahbubiya Girls' School in 1907, for which, in 1910, she was awarded the Kaisar-i-Hind Medal by the British monarchy. Their daughter was Sheila Walker (b. 1889), one of Begum Sarbuland's friends. Pernau, "Schools for Muslim Girls," 265.

17. Given in English, with translation.

18. All of her children, that is, except for the three elder boys already mentioned.

19. The reference is unclear.

20. Maharaja Kishen Pershad (1864–1940) was the prime minister of Hyderabad between 1901 and 1912 and again from 1926 to 1937. He published at least thirteen travelogues in Urdu over a thirty-five-year period under his nom de plume, Shad. For a biography by a contemporary, see M. Khan, *Maharaja Kishen Parshad*.

21. Maharaja Kishen Pershad had seven wives, Hindu and Muslim. Ghausia Begum was reportedly his favorite.

22. Better known today as the Yusufain shrines, this is a prominent landmark in the city. She visits the shrines before departure both as a sign of respect and to gain the saints' blessings and protection.

23. "Nafl tahiyyat al-masjid": prayers made when entering a mosque.

24. The shrine of a nineteenth-century Chishti Sufi.

25. The shrine of Baba Sharaf ud-Din is located just outside the city of Hyderabad. It now borders the city's airport.

26. This was probably Haidar Jivan Beg, whose title was Jivan Yar Jung (b. 1880). He studied at Cambridge and later succeeded Nawab Sarbuland as Supreme Court justice. His wife was Rashida Sultan. *Hyderabad State*, 17.

27. Begum Sarbuland's elder brother Nawab Zulqadar Jung Bahadur (d. 1951) was a judge in the Hyderabad High Court and later a state secretary. *Hyderabad State*, 35.

28. This refers to Nawab Agha Mirza, whom she calls here "bhai abba jan." He had left Hyderabad in 1897 and was then living in Lucknow.

29. Maulvi Abd ul-Bari Firangi Mahalli (1878–1926) was a prominent scholar and Sufi from Lucknow. His disciples were located across India. He communicated with them by mail. See Minault, *Khilafat*, 32–38; Robinson, *Ulama of Farangi Mahall*, 151–76.

30. Dame Una O'Dwyer was, from 1919, a Dame Commander of the Order of the British Empire and is remembered for having founded several charitable organizations in India. Her husband was acting Resident Michael O'Dwyer (1864–1940), who held a variety of high-ranking positions in colonial India. He is best remembered for having been lieutenant governor of Punjab during the Jallianwala Bagh massacre of 1919 and for having approved of the events of that day. Those infamous events of course postdate the exchanges described here.

31. This refers to Rashida Begum and Iskandar Begum, the latter of whom is mentioned above.

32. Begumpet Station would have been the train's second stop in Hyderabad. It is located about six kilometers from Hyderabad Station. Her use of this station meant that the five hundred well-wishers may not have ever known that Begum Sarbuland had joined her husband on the trip at all.

33. Mirza Akbar Beg (b. 1893) worked as an engineer in the Hyderabad government from 1908. *Hyderabad State*, 21. He was the father of Razia Sultan, who would marry Begum Sarbuland's youngest son, Halimullah.

34. Banda Navaz Gesu Daraz (1321–1422) was a Chishti Sufi saint. His shrine is among the most important in India.

35. *Subedar*: A company commander in the British Indian army. This post was the highest that could be occupied by an Indian within the British forces.

36. This likely refers to Gulbarga Fort and the attached Jama' Masjid. The complex was built by Bahmani and Adil Shahi rulers in the fourteenth and fifteenth centuries. They are now located well within the city's limits.

37. From here Begum Sarbuland switches to the Gregorian calendar. The transition is seemingly not random but reflects her departure from the state of Hyderabad and entry into the British-run Bombay Presidency. She maintains use of the Gregorian calendar exclusively from here until the end of the account, apart from a few days while on hajj in Mecca.

38. Mahmud Ahmad meets the couple again in Jeddah later in the account.

39. Located directly on the harbor, the Taj Mahal Hotel was then the pinnacle of luxury in Bombay. It was built by Jamsetji Tata in 1903.

40. Hamidullah describes his difficulties that afternoon in his own travel account. "There was no one in any of the agencies' offices in Bombay who were able to give me clear information on the present condition of the new Medina railway or to confirm if the Hejaz Railway

was running or if it had been blocked for some reason. I did not think it wise to spend too much time on this, so I put my faith in God, and within a few hours all the arrangements were made, and I boarded the P&O ship *Salsette*. As a precaution, I bought a ticket up to Port Said." H. Khan, *Safarnama-e Madina*, 1. He uses the word "I" here, not "we," despite the fact that he boarded the ship with his wife and her servant Amina bi!

41. Begum Sarbuland here actually uses the term "station" rather than port, perhaps reflecting her greater familiarity with railway terminology.

42. In Mecca, the Sarbulands were joined by a large party of family members and household staff who traveled there directly from Bombay. Their earlier departure may have been designed to let them travel on a newer, more luxurious vessel and to experience the new Hejaz Railway.

CHAPTER 2

BOMBAY TO SUEZ

بمبئی تا سوئز

On the Beautiful Ship

The name of this steamer is the *Salsette*. It belongs to the P&O Company.[1] It is beautiful and comfortable. The lights are all twinkling now, even though it is still daytime (fig. 2.1). Many lunch tables have been set out. The dining room is large and beautiful. We took the ship in as we walked to our cabin. I removed my burqa and gazed through the porthole out at the sea with wonder. The ship was still anchored. An English waiter arrived and said, "Madame, lunch will be at two thirty." I answered, "Yes." I know English well enough to be able to make myself understood easily. Then I laid on the bed and had Amina bi carefully arrange the luggage.[2]

Interesting Aspects of the Ship

Soon Nawab Sarbuland Jung arrived and said, "Come, the table is ready."

"Should I eat with everyone?"[3]

"Yes, come. Who will see you here, anyhow?"

"All right, let's go, then."

I put on a cloak that I had had stitched especially for this trip and covered my face with a Turkish niqab, and we went to the drawing room. Despite [my niqab], I still felt uncomfortable and shy. By chance, though, our table was completely separate from the rest. Only Nawab sahib could see my face; everyone else was seated at tables directly behind me. Once we were seated, I flipped the niqab back over my head. There was no one at the

Fig. 2.1 Begum Sarbuland traveled from Bombay to Aden on the modern and swift *Salsette* in 1909. Photograph: Green, Allan C. *Source: State Library Victoria.*

table but us. The waiter who served us was European. He was very excited and said, "Oh, I am serve Indian lady!"[4] It's quite droll that he would be so excited to serve me.

I returned to the room after eating to rinse my mouth and wash my face. Nawab sahib said, "Let's go up to the drawing room. We can sit on the deck and watch the sea." "You go ahead," I said. "I'll join you soon." He went up, and I followed soon after wearing my cloak and niqab. On deck there were several ladies reading books or strolling about. I went straight out, sat on a bench, and began to read a book. I read for a while and then got out the diary that I started keeping while still in Hyderabad. Soon it was time for tea. Nawab sahib came and said, "Let's have tea."

"I'll have tea in my room."

"Sure. I'll have it sent."

Everyone went for tea, and I went to my room. I took off my niqab. The waiter brought the tea. I drank it and lay down in bed to read my book. At evening time, the electric lights were illuminated. At seven Nawab sahib came and said, "You should get ready. I will change my clothes. Dinner is in thirty minutes."

"Yes, do get ready. I already am."

We both went to the table and sat in our places. The food on this ship is excellent. After dinner we went to the deck and strolled for an hour before returning to our cabin. Then we prayed isha and went to sleep.

First Morning at Sea

November 15, 1909: In the morning, we prayed *fajr*. Then we took a "sea bath," which made us very clean and soft. By then it was seven and breakfast was ready, so we went down to eat. I liked the mutton chop, so I had that. Then we had tea, and I went to sit in the drawing room. I spoke to the memsahibs for a while and then wrote in my diary as usual. Nawab sahib and the others strolled about. At eleven, the waiter brought beef tea. The ladies and I each had a glass. Then I went back to my room to write a few letters until mealtime. Later Nawab sahib came, and we went down to lunch. We heard that another ship would pass ours. Everyone stood up from their tables to watch. It came closer and closer until the passengers on that ship were looking at us and we were looking at them. Once this spectacle concluded, we ate, and I returned to the cabin.

Taking a stewardess with me, I went to the kitchen and made a tour of all the places on board. I saw every machine on the ship. The tableware is washed in a wondrous way. First, they put the dishes in a basin. Then they turn the machine [handle], and the dishes go up and down while the machine is filled with hot water. And that's it: the dishes are clean!

After the tour I went up to the deck, where the mems were waiting for me. Today I became friends with Mrs. Shane. We spoke for a long time. Then I wrote in my diary till teatime. Nawab Sarbuland Jung sahib and I went to tea and then returned to our cabin. After the asr and maghrib prayers, we went to dinner as usual. Then we strolled the deck, prayed isha, and went to bed.

November 16, 1909: I went to the drawing room after completing my usual morning routine. There I had a long debate with a mem about religion. She criticized Hindustani customs and practices. I did not hold back either. Then we began to joke with one another and moved on. We played a few games. The memsahiba asked me for some cardamom. "Give it to us," she said. I gave her a few pods. She noted down my address and also gave me hers. Suddenly Mrs. Shane appeared. She is elderly and has taken quite a shine to me. She lives in Allahabad but is going to London for a while. She asks me often about Hyderabad: "What is Hyderabad like?" she asks. I described the flood to her and praised my city so much that she grew eager to visit Hyderabad Deccan herself.[5]

Then it was lunchtime. I ate and went to my cabin. Nawab sahib did the same. We began to chat. Nawab sahib said, "What should we do? Do you want to go to Jerusalem or Egypt? We will have to wait for a week at Suez. We can visit either of these places in that time." We'll see what we end up doing.

We were still discussing this when suddenly there was such a tumult on the ship that I became frightened. Nawab sahib said, "Don't be scared. This is just a drill of what to do if the ship is caught in a storm." All the poor workers had to drop whatever they were doing and run to the muster. *Al-hamdulillah*, the drill soon ended. The day passed in conversation. After dinner we prayed and slept.

November 17, 1909: I woke up early and took a sea bath. I take a sea bath every day because I hear it is good for one's health. After breakfast Nawab sahib came and said that we will stop at Aden today. "We need to transfer from the *Salsette* to another steamer. Please get your things packed. We will disembark after lunch." "Very well," I replied. Amina bi and I got everything in order. Now it is 11:00 a.m., and I am on the deck, packed, ready, and waiting.

Soon I could see land taking shape in the distance. My memsahiba friends and Mrs. Shane were also all packed, and we were taking in the mountains and the sea. Then Mrs. Shane, Mrs. Koch, and I went into the drawing room. Mrs. Koch began to play music. It's strange. This is my first sea voyage, but I haven't felt dizzy at all. Nawab sahib and I have moved about in total comfort.

When she finished playing, Mrs. Koch asked me, "Do you also know how to play music?"

"Yes," I replied, "I know a bit of Hindustani music. In English, I can only play the scales and a few sharps. That's all."

Some of the ladies pressed me to play. "Please, play something!" I looked about first to see where the men were, but they had all gone out on the deck. So I sat at the piano and played the *ghazal* "Saqiya Mai Deh ki Ma Durdi Kash-e Maikhana-im."[6] The tune of this ghazal is very sophisticated, and everyone enjoyed it. Then Mr. King suddenly appeared, so I shut the piano and left.

Arrival in Aden

By then the ship was very close to shore. It was lunchtime. When the bugle sounded, everyone went to their tables. Just as we were sitting down, the ship dropped anchor. The ship came to a stop. We had arrived at Aden. We began to hear the buzz of sailors and passengers outside. After lunch, we went out on the deck to watch the spectacle. Before us we could plainly see Aden's shops, buildings, streets, cars, horses, camel carts, etc. We saw people navigating the sea in tiny little boats, some of them propelled by steam, others by hand-drawn paddles. Some of the boats moved with a broad white sheet that looked like a banner. They all zipped about so quickly; it was astonishing. They bobbed with the waves as they went, but the people inside them were at ease, as though they felt no fear at all.

Several boats approached our ship on either side. Inside were Habshis,[7] who dance about and make a ruckus. All the ship's passengers threw coins from the deck as a reward. I threw six *anas* myself.[8] I have heard that these savages speak Arabic. A month before leaving for this trip, I had hired an Arab woman as a servant so that I could practice spoken Arabic with her.[9] Thus, I had begun to be able to understand Arabic. I said a few things to them in Arabic, which shocked my English lady friends. They asked me, "Oh, what are you say?"[10] I replied, "This is our pure Arabic language. I speak a bit of it."

Then we all went up to the deck and sat in the drawing room. Notice was given that at 3:00 p.m. we would have to transfer from the *Salsette* to the *China*, which was larger.[11] Nawab sahib came at 2:00 p.m., and we went to our cabin and checked the luggage. We had two stewards on board, one who served our meals and another who looked after our cabin. We gave tips to both: two rupees for the first, four for the second. There is a third servant who keeps the bathroom warm. We gave him one rupee.

When everything was ready, Nawab sahib went upstairs while I took a nap in the cabin. When I awoke it was 3:30 p.m., not 3:00. Afraid [I was late], I rushed outside only to discover that everyone was going to tea. My sahib and I went too. Then it was five. We went to disembark. As is standard, we boarded a large boat that came to collect us. This boat ferried us to the *China*. We were assigned a corner cabin that I did not like. Nawab sahib had us switched; the new one was very nice. The bathroom and so forth are nearby. Amina and I organized the luggage. I washed up and changed my clothes before donning a niqab and cloak to go up on deck. Oho! This entire ship is filled with sahibs and memsahibs, all of them strolling about. My sahib and I too had a stroll. When my lady friends arrived, Nawab sahib walked alone while I went to speak to the ladies. The dinner bugle sounded. We went to eat. Then, as usual, we strolled on the deck before returning to the cabin.

November 18, 1909: After finishing our morning routine, we had breakfast as usual, then went to the deck. Nawab sahib spoke to some sahibs while I sat on a bench and talked with Mrs. Shane. Two girls of about six or seven came toward me. They were frightened by my niqab. I flipped the niqab back and called out to them. They ran over, sat close beside me, and began to chat.

Girl: "Who are you? Are you a pussy[cat]? Is that why you're wearing clothes like this?"

The word pussy[cat] made me laugh out loud. Mrs. Shane too was beside herself with laughter. I told them, "No, I am not a pussy[cat]." Their fear evaporated. They gave me their parents' names and told me about their homes. One of them said, "I have a beautiful doll." The other said, "I have a beautiful little diamond brooch, a pair of earrings, and three silver bracelets." I really enjoyed listening to them. Eventually their parents arrived and began to laugh on seeing them so engrossed in the conversation. The girls had me write their parents' names in my diary. Their names are George Webb and Mrs. Webb.

Nearby there were a few ladies playing cricket. It was nearly one. I went to the cabin, got ready for lunch, and went to the dining room. I returned to the cabin after eating. I changed and laid down. At four, I washed up, did *vuzu*, and performed the asr prayer. Then I got dressed, put on a niqab, and went for tea. Nawab sahib was there. Afterward I went to the cargo-loading area and inspected the machinery. Everything is done so precisely. Back on deck, I chatted with Mrs. George Webb. I also brushed up on my Arabic until the two girls reappeared and we began to talk.

Mrs. George Webb said, "Come, let's take a walk before going to eat."

We strolled around. The dinner bugle sounded. As usual, after dinner I went back to the deck. Nawab sahib was strolling around on the other side of the deck. Mrs. George Webb then came to meet me again. We began to chat.

"Would you like to go to the barbershop tomorrow?" she asked.

"There's a shop here?"

"There's a barbershop at the far end of the ship. You can buy brooches and ribbons and such. Whatever you want, you will find it there."

"Very well. If you are going, I will join you. I'll see what all they have there."

Then I said good night and retired to my cabin. Nawab sahib was praying isha. I prayed too.

November 19, 1909:[12] As usual, I woke up, had a sea bath, got dressed, and had tea. Then I went straight to the deck, where the first person I met was Mrs. Webb herself. The moment she saw me, she reached out and flipped the niqab off my head.

"What's happened to you? Don't put this over your face like that."

"Madame, it is our customary practice [to wear the niqab]. Please, leave it be."

Even still, she wouldn't listen. I left the niqab open for a while before eventually lowering it again. Later, I went with the ladies to look out at the sea. The lunch bugle sounded. After lunch I went to the cabin and took out two pounds and a few shillings. Back on the deck, I saw Mrs. King and spoke to her for a while. Mrs. Webb and Mrs. Shane came over too.

"We are playing a game," they said. "The game is to guess how many miles the ship travels in an hour. Whoever is correct wins the prize. You should wager a shilling. I'll enter it for you."

"Take the shilling, and enter it for me. I don't understand this game at all."

I gave her a shilling to humor her, but immediately after, we learned that a sahib had won. He had won the previous thirteen days too. Mrs. Webb and I went to the barbershop. The barber had decorated it very nicely. He was doing someone's hair when we arrived. We looked at the products for sale. Mrs. Webb handed me a book and said, "Can you read this? It is in English."

"Yes, I can read it," I said, and we began to read together. She sat on one chair; I, on another. She began to knit a woolen stocking. I took it from her and said that I would knit while she read. Eventually, the barber finished their work, and we returned to the shop. I bought a brooch, a few safety pins, and some postcards with a picture of the *China* steamer. Then we left, and I went to write the postcards. I sent one to each of my children in Hyderabad. Letters can be mailed from the ship itself. They will be forwarded to the post when we reach Port Said. Anyhow, I wrote the postcards and mailed them. When Nawab sahib came, we prayed asr and then went up to the deck to watch the sea. A piano has been put out on the deck.

"There will be a dance tonight. You must come," Mrs. Webb said.

"I will certainly come," I replied.

After dinner we went to the cabin and prayed isha before going on deck. I sat with the ladies while Nawab sahib strolled with the sahibs. The music began. Nawab sahib said, "I'll go to the cabin. I want to pray some *vazifas*."

"Go ahead. I'll watch the dance and come later."

Then I went to sit with Mrs. Shane. The sahibs walked over to ask the memsahibs, "Will you dance with me?" Most of the mems refused, and most of them also got ready to dance. The dance began. A steward played the piano; everyone enjoyed his playing. There were many different types of dances. I had never seen an English dance before. This was my first, and I took it all in. Mrs. Shane stayed with me. She really wanted to dance with someone. This was completely astonishing to me, given the fact that she was nearly sixty years old. She herself told me that she was sixty. Imagine wanting to dance at such an age! Sadly, her desire went unrequited. No sahib asked her to dance. This greatly disappointed her. Thank God, the event finally concluded at 11:30 p.m. I bid her good night and went to my cabin. Nawab sahib had already finished his vazifas.

November 20, 1909: I woke up early, had a sea bath, and then we went to breakfast. I said to Nawab sahib, "Please have my trunk brought up [from the hold]. I want to retrieve some clothes. The luggage hall is open today." My sahib had it brought. I took a few clothes out and put some others in. I closed it back up and had it returned to the luggage hall.

Back on the deck, I wrote letters to Lady Walker, Mother, Khurshid, and Father before meeting with my memsahib friends. After lunch I lay down in my cabin for a while. Nawab sahib returned after getting a haircut at the barbershop. We bathed and went to tea. Afterwards we strolled the deck. I spoke to a few ladies who were there. At maghrib I went to the cabin to pray and then returned again. Nawab sahib was doing some writing in the drawing room. I had a stroll and ran into Mrs. Lyall and Mrs. Webb. There was still a quarter hour until dinner. I told Nawab sahib to get ready. He went to the cabin while I continued to walk. After dinner I prayed isha and went to the deck again. Mrs. Webb took my hand, and we walked together before settling into the deck chairs.

"How do mem women style their hair?" I asked.

"Come to my cabin. I will show you."

We went to her cabin. She let down her hair and then showed me how she did it up.

"Now you do yours," she said.

"My clothing is different from yours. That hairstyle won't look good on me."

But she would hear none of it. She did up my hair and said, "Let's show Mrs. Lyall."

"You'll have to ask Mrs. Lyall to come here. I won't go outside like this."

Mrs. Lyall came, and we all laughed together for a long while. Then I let my hair back down and returned to my cabin.

November 21, 1909: I woke up and prayed fajr as usual. Then I had a sea bath, got dressed, and had breakfast before going on deck. I wrote a letter to [my aunt] Khala Masudah. Mrs. Studes came and said, "Please read your diary aloud to me. What is it that you write every day?" I read it out to her and Mrs. Shane in translation. Al-hamdulillah, my translation from Urdu to English was good enough that they could both understand and enjoy it. When we were done, they stood up to leave because it was time for church, for today is Sunday.

They all left until only Mrs. Shane remained with me. "Won't you go to worship [*namaz*]?" I asked her.

She replied, "No, no, I will not. I will stay right here and talk with you."

She sat down beside me and wrote out a letter to her son, who is in Shimla, while I wrote in my diary. When she finished, she said, "Let's to go the barbershop to buy stamps. We're going to pass Suez at one today. We should have all our letters posted."

"All right," I replied, "I'll post my letters too."

We gathered up the letters and took them with us. We put stamps on them and posted them.

A View of Suez

Suez slowly came into view around eleven. Some of the sahibs were at church, while others were on the deck reading books or perusing the newspaper. Then we reached Suez. It seems to be a beautiful city. The buildings are nice. It's verdant too. Aden was completely dry, but in Suez I saw a lot of greenery. Everyone here appears to be Arab. Their clothes are Arab. Everyone is Muslim. Many Arabs came out in their little boats and tied them to the ship. They climbed up on to the ship bearing large baskets and laid out their wares. They had all kinds of things. The mems and the sahibs began to shop. I myself bought three necklaces, a pair of bangles, and a black shawl with excellent embroidery. I also said a few words to them in Arabic, which made them happy. They said, "You are Islam. All these people are Kafir. You are good. Al-hamdulillah, *masha* Allah, Muslim.[13] We will give you a cheap price. *Wallahi*, we will be honest with you." They went on displaying and selling their products for two hours. So many people bought something or other. Finally, it ended. They climbed back down and untied their boats. They said "as-salamu alaikum" and "marhaba marhaba" as they left. Everyone here gets about in these little boats. They carry people to shore for thirty pence and then bring them back to the ship too.

It was time to leave. The ship set off again. From here our path was very strange. There is dry land on either side, right up against the edges of the ship. The ship travels between these two banks. Suez can be seen from right up close. It looks green and prosperous. You can even see trains passing. Stations appear regularly. You see laborers at work. It has been suggested that the canal be expanded so that two ships may pass in either direction at once. It is already 5:30 p.m., but the laborers are still at work. The ship is proceeding so slowly you cannot discern if it is stationary or in motion. They say this place was once completely dry; it was all land. A ship could not pass through here. Travelers would get down [at Suez] and travel by train. Water was only brought here and made deep after digging endlessly. It cost hundreds of thousands of rupees to bring water here and to make it deep enough for it to become a sea. Now ships pass easily. Construction is ongoing. This place is called the Suez Canal. It is very long.

The dinner bugle sounded. We could still see everything. We had dinner and then returned to our cabin.

Port Said

November 22, 1909: We woke up as usual, got dressed, and went to breakfast. We can still see land. We heard we would arrive at Port Said soon, and at 8:30, we did. Nawab sahib said, "Get your things together quickly. We must disembark here." I hurried to the cabin and prepared everything. I told Amina bi to get ready, and then I went on deck. The ship had already stopped moving. The Arabs again brought their wares to sell. The passengers began to make purchases. I bought two *muzak* brooches and spoke to the vendors briefly in Arabic.[14] As usual, Nawab sahib tipped the steward, who was European, half a pound. He gave four rupees to the stoke, and two rupees to the sweeper.[15] To our waiter he gave three rupees. I said, "With your permission, I will go and see Port Said with Mrs. Shane" (fig. 2.2).

By Nawab sahib's leave, I put on my cloak and covered my face with a niqab. I collected a few pounds and shillings and boarded a small boat for the ride to shore with Mrs. Shane. To disembark from the ship, a ladder

Fig. 2.2 A photochrom of the quay and a market street in Port Said, circa 1900–1905. The city would have looked similar when Begum Sarbuland visited. *Source: Library of Congress, Prints & Photographs Division, LOT 14192, no. 227.*

is placed in the boat. You climb down this ladder and sit in the little boat, which bears you to shore. None of the larger boats that unload passengers came to the ship. These smaller boats charge you a fee to come. The fare from ship to shore is thirty pence. Well, Mrs. Shane and I reached the shore. Al-hamdulillah, everyone here is Muslim. They wear Turkish fezzes. The people here have such fair skin, though quite a few are darker too. Their dress is largely *angrezi*, though some wear a long kurta with a coat and a fez. All the city's residents speak Arabic and English. Their English is broken. They don't know Urdu at all. They speak French better than English. There is a Jewish community here too. What I liked most is that Islam is given great value. Muslims are highly respected. Mrs. Shane said, "These Arabs are so brave and daring. There are hardly any angrez people here. I am afraid that if we go into the city, they may do something to us."

"No, you should go," I replied. "You are with me, a Muslim."

When the memsahib and I reached the main gate of the city, a few policemen salaamed when they saw me and asked, "Are you Turkish and Muslim?" I returned their salaam and said, "Al-hamdulillah, I am Muslim." They were glad to hear this.

"Marhaba," they answered. "What brings you here?"

"I am making a pilgrimage to Mecca."

"God's mercy be [on the Prophet]."

Mrs. Shane asked, "What are these people saying?"

"They're asking where we have come from and where we are going."

One of the men began to walk with us and said, "I will show you the way." I told Mrs. Shane, "Let's let him come with us. It's OK. He will guide us." Taking him along, we first went to the Cook office.[16] This is where you buy tickets and such. Mrs. Shane said, "Your sahib promised to come to the Cook office [to meet us.] We should wait here for a while. When he comes, we can explore the city."

I replied, "All right."

Nawab sahib came a short while later and said, "Go tour the city with the memsahib, but be sure to come back by four. We need to board another ship. First, we'll board a P&O ship and have lunch there. Then we'll change ships again."

"All right. I will go to the ship by two p.m., inshallah."

Then I went into the city with Mrs. Shane and our escort. Mrs. Shane was very frightened. She said, "Tell this man to please go away. We will manage from here."

But he replied, "No, please let me accompany you. I will arrange everything for you."

I said to her, "Please don't be afraid. Let him accompany us. He will help us with our tasks, and then we will give him a tip."

The two of us went into a shop. Mrs. Shane asked him repeatedly, "Where are you taking us? Please take us somewhere where there will be other memsahibs."

This made me laugh. "Look," I said, "there are so many English people in this shop. You can relax now."

We went in. The owners are Greek and speak English and French. When they saw me, they said "marhaba" in Arabic and then spoke to the memsahib in English. They gave us chairs to sit on and brought out a lot of merchandise. I bought a jacket, a blouse, a coat, a warm Turkish kurta, two pairs of warm stockings, and a pair of shoes. I didn't find any warm cloth that was appropriate for a sari. I had everything packed up and asked them to keep it for me. "We will collect it on our way back," I said. "We'll continue shopping for now."

Then we went to a doctor's clinic. Mrs. Shane bought a purgative. Then we visited a fruit vendor. One fruit caught my attention, but when we peeled it, it was still unripe. I did not like the taste. We bought some dates and then went to another shop. Here we saw two Egyptian women wearing jewelry and other ornaments. Their burqas were very strange. They had a round, golden ornament at the nose. Their eyes were completely open, but their mouths were entirely covered. Below their burqas they wore blue kurtas with frills and other designs. Their hands were covered with henna, and they had golden bangles on their wrists. They were buying some cloth. They looked at me, and I looked at them. Each of us was wondering at what the other was wearing. Their kurtas were so long that I was unable to see their pants. I saw curly hair peeking out [from their burqas] at their temples. These women had a good look at me and realized that I might be Muslim.

One of these bibis came up to me and asked, "Are you a Muslim?"

"I am," I replied.

She kissed my hand and said, "You know Arabic."[17]

I replied, "No, I only know a little. I know Hindi."[18]

"Wonderful. You sit beside me."[19]

I went and sat beside her. This bibi's feet were tattooed. The practice of tattooing exists among the Hindu women of Hyderabad. I asked the bibi, "What is this?"[20] She replied, "Look this. Good."[21] I didn't understand the rest of what she said, except that her people consider these tattoos beautiful and that they tattoo their hands too. It is a pity that so many questions worth asking went unasked because our languages are different. Mrs. Shane too stared at these bibis in astonishment. The bibis said to her, "Come here, servant girl, come here."[22] But she didn't understand.

I bought a belt and some cloth for a warm sari, said goodbye to the women, and left. Then we went back to the first shop to retrieve our purchases. I still owed the shopkeeper a pound, but we had already spent all our money. He sent a boy with us. "Take this boy with you and give him the pound."

On the Same Ship Again

We returned to the ship around twelve thirty. We were both famished. The policeman carried our items to the dock. I gave him a shilling, which made him very happy. He then asked the memsahib for a reward. She said to him, "Oh, the bibi gave you a lot. You go away now." But he did not go. When

I boarded the boat, he held my hand to steady me. Then he gave a salaam and left. Back on the ship, Mrs. Shane came to my cabin. We opened the packages because our items had all been wrapped up together. She took her things and gave me mine. We counted the items and ensured that everything was there and then gave the boy the pound we owed. We had Amina bi take Mrs. Shane's items to her cabin.

Nawab sahib returned and quickly got his luggage together before going to the table. After lunch he gave our baggage to the Cook representative, who had it transferred into a boat. Then my sahib and I went to the deck to meet with everyone. I spoke with all the people I had met while at sea. All the memsahibs who had become friends with me during these few days together expressed their sadness at my departure and shook my hand affectionately. There was a magician performing, and everyone was engrossed in his act. I watched for a few minutes, too, before turning away. I met Nawab sahib on the way, and we went to the boat together.

When we got to shore, I sat in the boat while he went to the Cook office. He was gone for so long that I began to get bored.[23] I left the boat and began to walk around. Suddenly two Muslim men walked by speaking in Urdu. This was the first time that I had heard a conversation in Urdu since leaving Bombay. I was excited to hear it. I asked them where they were from. "We are from Punjab. We've come here to look for work." Then some Arabs walked by. I chatted with them in Arabic. When Nawab sahib returned, I got back into the boat with Amina bi.

On the Ship to Jerusalem

From here we boarded another ship headed for Jaffa, Haifa, and Jerusalem.[24] The name of the ship is the [*Imperatrice*]. The boat left at six. This German vessel is so dirty that my mood was completely spoiled.[25] Our cabin is large, with four beds, but I do not like it. Dinner here is at seven. We went to eat. At dinner there were a few priests and ship officers. The meal was just a few cuts of foul-tasting meats. There were no puddings etc. at all, though they did have oranges. I couldn't bring myself to eat this German food. When the "salad" came, it was just a leaf of spinach. Not good spinach, either. After eating we went on deck. The lounge area only has rattan couches. It is very uncomfortable. The area around the room is filthy too. There are poor Arabs and Turks lying about everywhere underneath the open sky. Some of the Arabs sell fruit. This area is known as third class. I had a good look around. Then Nawab sahib and I went to the uppermost deck and had a stroll. There are many deckchairs, but they are all dirty and wet. All the officers and passengers speak German, which I cannot understand at all. Men and women share the same restroom. I don't like that either. We have returned to our cabin.

Today is November 23, 1909. Right now it is six in the morning. We prayed the morning prayer, got dressed, and went to tea. The tea is very bad too. I only had milk with plain toast because there was no butter. At seven we reached a settlement called Khafa [Jaffa]. The ship stopped here for an hour.[26] I went onto the uppermost deck, found a dry seat, and sat to write in my diary. Nawab sahib came up, and we chatted. He told me that we will reach Haifa, which the Arabs call Kafa, tomorrow morning, inshallah. We will disembark there and go to see a place nearby called Muzik. We will also visit Nazareth and Tiberias. We were told that lunch is served at 11:00 a.m. I went and sat in the room with the rattan chairs.

An Encounter with a Poor Turkish Woman

After leaving the room I saw some poor Turkish women walking around. One of them came up to me. She spoke to me in Turkish; I understood nothing. How could I respond? Suddenly her brother appeared and spoke to me in English: "Do you speak English?"

"Yes," I replied.

He spoke to me in English and told me about his sister. "Look, do you see these marks on her face? She teaches sewing at a school. One day she was on her way to class when she was attacked by thieves. They beat her so badly that her entire face was covered in wounds. These scars are from those wounds." I felt very bad for her when I heard this story. These siblings are Muslims. I gave them a shilling, which made them very happy. They salaamed me and left.

Then the bugle sounded, announcing lunch. We went to the table. *Hae hae*, that same stinky food again! I didn't eat it. My sahib didn't like it either, but he knows a bit of German, and when he spoke to the officers, they encouraged him to eat it. As for me, I asked them "to bring me a boiled egg with toast." They brought it. I sprinkled some salt on it and ate it. Then I went back to my cabin and began to write in my diary. At seven we were served the same food as before, and again, I didn't eat it but had egg with toast. Then we returned to the cabin and, by God's grace, fell asleep.

Notes

1. The couple's choice of the vessel *Salsette* itself reflects Hamidullah's passion for the newest technological and logistical advances. The ship was then brand new and the fastest

in the P&O fleet. It began regular operations in late 1908. Between 1908 and 1914, the vessel shuttled between Bombay and Aden, breaking speed records for the journey. In 1917, less than a decade after its creation, the ship was traveling between London and Bombay when it was torpedoed by a German submarine in the English Channel. "Ship Fact Sheet—Salsette (1908)."

2. Her servant Amina bi was about ten years younger than Begum Sarbuland. Their relationship was quite close, and she remained a frequent visitor to the Sarbuland household until the end of her life. Her daughter Muhammadi Bi was herself employed by the family to assist Begum Sarbuland's daughter Qudsiya Begum. Amina bi continued to visit the family into the 1950s, and Begum Sarbuland's grandchildren, who called her "Anna Ma," still remember her affection. Interview with Najeeb Jung, May 23, 2022.

3. She questions if she should abandon her observation of purdah and take her meals in full view of unrelated men.

4. "O, ham Indian lady ko khana khilata hai."

5. In September 1908 the Musa River, which flows through Hyderabad, rose sixty feet. It washed away three bridges, caused massive flooding, and killed an estimated fifty thousand people. Reservoirs were later constructed to prevent future tragedies.

6. A Persian-language ghazal by Saadi Shirazi: "Give me wine, cupbearer, for we are the dreg-drinkers of the tavern."

7. The term *Habshi*, when used precisely, refers to people from the Horn of Africa. It is, however, more typically used in a general—and often derogatory—sense to refer to people of African descent. In this case, the people to whom Begum Sarbuland refers are Somali. Despite her excitement for meeting Muslims on this journey, she does not at all indicate a willingness to acknowledge that these Somalis are Muslim.

8. There are sixteen *anas* in a rupee.

9. The Arabic vocabulary she uses in the diary suggests that this servant came from Arabia.

10. "Tum kya bolta hai?" The reported speech of Begum Sarbuland's English interlocutors is typically grammatically incorrect. It is unclear if she generally conversed with the women in Urdu or English or both. Her seeming mockery of these women's Urdu is a striking reversal of Euro-American travel writing about India, which similarly renders Indian speech as ungrammatical.

11. Another P&O vessel, built in 1896.

12. The text erroneously gives the seventeenth.

13. "Enti Islam. Hada kulluhum kafir. Enti tayyab. Al-hamdulillah masha Allah, Musulman." Begum Sarbuland often includes spoken Arabic phrases in her diary without translation. The Arabic she records is typically an ungrammatical reconstruction of what the speakers might have said. It sometimes includes Urdu words that are not used in Arabic. In most cases, though, the meaning is clear, and it is a testament to her acuity and commitment that she learned so much Arabic so quickly during this journey. I reproduce the Arabic text here without correction, and my translations are necessarily approximate.

14. The meaning of *muzak* is unclear.

15. What "stoke" refers to is unclear. It could possibly refer to the category of steward mentioned earlier who "keeps the bathroom warm."

16. This refers to the local representative of the well-known travel company Thomas Cook. Begun in England in the 1840s, it had become a popular tour operator both within and outside Europe by the end of the century. It was particularly popular among wealthy Indians, who relied on it to arrange their tours abroad. For an enlightening study of the company's history, see Withey, *Grand Tours*.

17. "Enti 'araf 'arabi."

18. "La, ana 'araf shuaya. Ana 'araf Hindi."

19. "Zain. Enti ajlas ila 'andi."

20. "Mish hada?"

21. "Hada shuf tayyab."

22. "Enti tatal, kaniz, tatal."

23. He had gone to arrange for passports. "We landed at Port Said and assumed the incognito of ordinary Indian pilgrims. . . . The Indian pilgrim, who had previously obtained a passport in India; but who out of foresight obtained another at Port Said, had very little difficulty in obtaining it, the whole process taking a couple of hours." H. Khan, *Pilgrimage to Mecca*, 4.

24. She calls these places Jaffa, Hafa, and Bait ul-Muqaddas. Curiously, she never actually visits Jerusalem. Her writing on Palestine, Jordan, and Medina here can be compared with several pieces by her husband, Hamidullah, who wrote in detail about these portions of the trip. His writing is found in the second appendix.

25. The ship was in fact Austrian, belonging to the Österreichischer Lloyd company. H. Khan, *Safarnama-e Madina*, 2. She gives the ship's name as *Almafi Tahter*, but Hamidullah calls it the *Imperatrice*. H. Khan, "Pilgrimage to Medina I."

26. Assuming that Khafa is indeed Jaffa (which concurs with Hamidullah's account), he states that the ship remained there "all day" loading and unloading cargo. He writes about the city at length. H. Khan, 2–3.

CHAPTER 3
PALESTINE
فلسطين

Arrival in Haifa

November 24: We woke up very early and prayed fajr. Then we went to tea, but I did not have any. We have arrived in Haifa, and the ship has stopped. Nawab sahib, Amina bi, and I disembarked and boarded a boat for shore. While still aboard, we entrusted our luggage to the Cook representative. It was then brought by him. It took about an hour and a half to arrive. The ship's officers traveled to the shore with us. This city is beautiful. There are many Arabs here. Their complexion is rosy, like that of Europeans. They are made even more beautiful by the celestial light of the Prophet Muhammad's faith.

I went and stood some distance away from the main thoroughfare. "You wait here," Nawab sahib told me. "I'll be right back. I'll come and get you."

"All right, I will wait here with the luggage. Go ahead."

After Nawab sahib left, I began to look around. The men here wear a coat, pants, and a Turkish fez. The language is Arabic. Those who are German speak German and English, though they don't speak it well. About fifteen minutes later, my sahib returned and said, "Let's go. Please come."

He took me to a nearby building. This was a German office. We arranged our tickets and such. There was an Arab boy standing in the room. I spoke a few words with him in Arabic. He seemed to be a very cheerful person. He brought a book and began to show me pictures in it while speaking rapidly in Arabic. Because I only know a little Arabic, I could only answer the few things I understood. When I couldn't understand him, I just laughed. Then I said to him, "I saw a bread maker's shop in the bazaar. There were kebabs and warm breads. Can you bring me four anas' worth of kebab and bread?" I spoke in Arabic with the greatest difficulty, until he understood me. Al-hamdulillah, he understood. He took four anas from me and came back with some kebabs and bread ten minutes later. I put the items in my tiffin basket and gave him four anas as a tip. This made him very happy. By then Nawab sahib had finished with the arrangements. After leaving our unnecessary luggage with the agents, we engaged a beautiful phaeton carriage. We got in, bringing the luggage we would need with us. We had Amina bi sit [up front] with the carriage driver and set off.

It was twelve noon. We were unable to see the whole city, but we passed by many of Haifa's historic places and buildings on the way. Then we were in the outskirts. The carriage was led by three horses. On the way, we passed some other travelers on the road. Some were on camels, some on horses, and some on foot. I was told it would take us four hours to reach Nazareth. Although there were very few lakes or wells along the road, the country has a lot of vegetation. While traveling, we encountered an Arab man who rode his horse alongside our carriage. Rural Arab men here wear a colorful, three-cornered cloth on their heads on which they place two close, round coils. But it suits them. They wear a long, loose robe, like a waistcoat, open at the chest such that it reveals the shirt beneath it. This is tied off at the waist, over the pants. Some, though, just wear a long kurta cinched at the waist. Others wear a type of pajama akin to a Punjabi *shalwar.* They wear these with just a tunic and a Turkish fez. The women only wear a dressing gown, something like a frock. On their heads they wear a scarf tied over their ears. Children wear the same type of frock. Their language is Arabic. Even the Christians speak Arabic. In fact, if the Christians here speak any English at all, it is broken and halting.

Around noon we found ourselves near a small settlement on a hill. We got down and had the carriage wait while we visited the village on foot. On the way up, we saw a wizened old man descending the mountain. As we reached the edge of the village, he advanced to meet us and greeted Nawab sahib with an effusive handshake. His clothes were entirely Arab in style, and his face was radiant. He began to speak to Nawab sahib in Arabic. All I understood was that on the top of the hill there was a smallish village and the shrine of a famous man, to whose family this man belonged. He had been among the great scholars of his time. I listened until this point and then began to walk again. On my return, I saw that the elderly man was still deep in conversation with Nawab sahib. After twenty minutes, he shook hands and left. We went back to the carriage. An hour later we saw a well where Arab women were busy drawing water. The driver had our water pouches filled, for which I gave the Arab women two anas. These Arab women didn't ask for anything; I gave it of my own accord. It made them happy. Our carriage driver is an Arab Muslim. The poor man shows and explains everything to us as we go.

One strange thing is that the Arab horseman who began to ride alongside us two hours ago was still with us. He also stopped at the well and gave water to his horse. Then he left with us again. Thirty minutes later our driver stopped at another well to have his meal. There were a few Arab women at this well too. The water didn't look very clean. The women were giving water and feed to their livestock. Our coachman loosened the horses' harnesses and gave them water and feed. I opened my tiffin and brought out some food, including biscuits, almond sweets, and the bread I bought in Haifa. We ate this food and fed Amina bi as well. We gave the leftover bread and kebabs to the coachman. I saw that the Arab horseman was still there, so I offered him some almond sweets. My mother had made these sweets in Hyderabad and sent them with me, but I hadn't needed them on board the ship. Now they were coming in handy. The horseman happily accepted them, saying, "I thank you."[1] He ate them and then said, "That was very good."[2] Then he said, "Where do you live?"[3] I replied, "I am on a tour of Mecca and Medina, after here."[4]

I gave some biscuits and sweets to the Arab women. They salaamed me and happily ate them. In return for the sweets, the Arab horseman gave sahib and me each a cigarette. "Smoke," he said.[5] I said, "I don't smoke."[6] I asked him to excuse me. Nawab sahib has brought a cigarette box with him in case he would like to give it to someone as a gift, but he does not smoke. I took a few cigarettes from it and gave them to the Arab horseman. "You smoke these Indian cigarettes."[7] He salaamed and happily accepted them.

Jesus's Birthplace

Our carriage was soon ready, and we departed. We saw buildings here and there as we went. There is a mountain here called Jabal Musa [Mount Moses], but they say that this is just its name; this is not Mount Sinai. That mountain is located on the route between Mecca and Egypt, they say. Even still, we prayed a fatiha for Moses when we heard the name. Once we passed Jabal Musa, a town came into view, with its buildings and churches. This town is called Nazareth. Jesus was born here.[8] Mary [Bibi Mariam] once lived here, and the place where she would worship God is located here. This is why many Christians choose to live here. There are many grand churches. The Christians here are very devout. However, Muslims do a lot of business here too. The Christians tend to stay in their own areas, so much that many of them become vexed if they travel into the city because there are so many Arabs, even though the Arabs never bother them. Despite this, the Christians are afraid of the Arabs.[9]

In the City of Nazareth

We have arrived in the city of Nazareth. The carriage halted at a German hotel. There was an Arab youth at the entrance. He salaamed us and began to unload our luggage. The owner, who was German but who spoke English, came out and brought us into the hotel along with our things. The hotel was beautiful and clean. The room we were given is excellent and comes with furniture. The sheets are white and clean. We arrived at three on the dot. Nawab sahib and I washed up, did vuzu, changed our clothes, and prayed zuhr. We got a separate room for Amina bi and stored our baggage in her room. The two of us had tea and placed an order for dinner. Then, umbrellas in hand, we left Amina bi with the luggage and went out to visit Nazareth and its famous sites. We first learned that there are many mission schools here. Children are given excellent instruction in English, German, and French. While we were out, I heard small children speaking in good Arabic in addition to these three languages. It made me sad to think of my own condition: even the Christians can speak Arabic, and yet I, a Muslim, cannot speak "the Arabic language, the language of Muhammad," as it is called. Some Christians can only speak Arabic and don't know any English at all.

A Visit to the Site Where Mary Worshiped

We went on touring the city until we reached Mary's worship place.[10] It was here that the spirit of Jesus, the "Spirit of God," entered her womb.[11] We went inside the building. It is very old. Outside there is a large gate, a courtyard, and some newer buildings. The old part is inside. Walking in, we were approached by one of the padres [priests], who are known by the name *bap* and are addressed as "father." They have renounced the world. They wear a small round hat and a coarse tunic beneath a thick, brown gown. They tie a rope of twisted threads about their waist. They carry a rosary [*tasbih*] and have large mustaches and long beards. Their hair is trimmed, and their faces are rosy. The padre came up to us and cordially shook our hands. Some of them are called "fathers" and others are called "brothers." What do age or rank matter when they have chosen to renounce the world? They become known as simply brother or father.

This father led us to a particular door and unlocked it with his key. We went inside with him. It was a beautiful room filled with images of Jesus and Mary and many important fathers, popes, and bishops besides. There was an image of Gabriel too. I can truly say without exaggeration that these images are a treasure. In our own time, I have never seen such beautiful and refined work as this. They are a sight to behold. In our religion, Islam, images of Gabriel are difficult to find. Impossible even. But here, these images are not controversial. My interest is in their artistry and beauty.

They have installed marble wherever the ground is flat enough. Everything seems very old. This place is on a mountain from which three connected rooms have been carved. A crack in the mountain above provides natural ventilation. The entire setting makes a profound impression. Ancient pottery has been placed here and there. It is as though this mountain was designed for these rooms. One of the rooms, I learned, was designated for worship, and the second for cooking. Mary would use the third as a sitting room. We prayed a fatiha for Jesus and for Bibi Mary and then continued our tour of this remarkable building.

A Temple of Idols near the Prayer Room

The father then took us into the next room, where we examined the idols and stone images that were worshiped in earlier times. This place has been excavated from the ground, along with all these idols and images. The ceiling and the walls are covered with designs of exquisite craftsmanship. The idols themselves are amazing. *Allah re!* They have been so finely sculpted that you can see even the individual stitches in their clothing. Their clothes, which have been carved into the stone, are something like a chador or a sari. Some of the idols are naked. But—*Allahu akbar!*—the craftsmanship and design are so fine that it seems as though they might speak aloud at any moment. The ancient pottery here appears to have been previously used. I was amazed by it all.

Jesus's Table

After a careful examination of these idols and vessels, the father led us to another building. Here we saw more fine images, as well as a massive rock the size of a boulder. This is called Jesus's Table. We were told that this is where Jesus used to eat with his twelve disciples, but only God knows the truth. The table was covered with a cloth. Opposite it was a wonderful painting of Jesus with his twelve disciples. All these buildings have been carved out from the mountain. In fact, they are not buildings at all but rather subterranean caverns. They are all well lit, and all are like blessed places. Our Christians do well to protect and maintain them.

Then we went outside again, and the father led us to a third place. Here we saw excavation pits being dug all around. A look into the pits revealed that they held stone houses that had long since collapsed. Now they are being excavated following clues from historical chronicles.

By now it was getting late. It was nearly time for the maghrib prayer, so we took our leave from the father. He asked us for some postage stamps from Hindustan, but, unfortunately, we had none. "We will send you some," we promised. We left and passed through the city, taking it all in until we stopped at a mosque and prayed maghrib. Then we returned to the hotel. The food was waiting for us. We sat to eat. The hotel owner's wife had cooked the food herself, and for that reason, it was very good. We ate until we were stuffed. Then we retired to our room. We gave Amina bi some money and told her to go and buy some food for herself. We went to bed at nine, after praying isha.

November [25], 1909[12]

I woke up early, prayed fajr, got ready, and went down to tea. We were served buttered toast and honey. I ate a lot. Then we went out to see the sights. We saw more historical things today.[13] They have excellent olive tree products here. I went into one shop to have a look, but Nawab sahib said, "Keep looking. We will buy from wherever we find the best products." "All right," I replied. We continued on. Two women showed us some handmade collars and scarves. I liked them, so I bought two collars and two scarves for nine francs. Then we returned to the hotel.

Departure for Tiberias

By then it was eleven. I washed up and went to the drawing room to eat. Back in the room, I packed up our things and told Amina bi, "Go and have the carriage readied. We will be departing for Tiberias." We were to take the same carriage that had brought us from Haifa. We paid our bill, climbed in the carriage, and asked the coachman to take us to Tiberias. The journey from Nazareth takes four hours. We stopped to visit three more ancient buildings all along the way. I went outside after I had finished visiting the final historical site. Nawab sahib was still visiting the site. Outside, I happened to meet a Christian bibi. She did not speak English or anything other than Arabic. I understood a bit of her Arabic. The gist of it was, "My house is near here. Come visit and buy something from me."

Once I had understood her meaning, I went along with her. She took me to a small room that was on the verge of collapse. She showed me a ladies' collar she had made by hand. "Buy this for a *majidi*."[14]

"I've already bought one like this," I said. "I don't want this."[15]

Meanwhile, Nawab sahib had finished his visit. When he saw I was not waiting outside, he came over to the house to find me. "I was worried you had gotten lost. Come on, let's go, hurry."

We got back into the carriage, but the bibi followed us.

"OK, have it for half a majidi."

I didn't need it, so I declined, and we set off. Soon the metaled road ended and became very rough. I had the carriage stop twice on the way, once for asr and again for maghrib.

Now it is 7:00 p.m. We are in Tiberias. We're staying in a hotel again. After our luggage was brought in, I washed up and did vuzu. Then, I got dressed. Dinner was already ready when we arrived. We shared a table with three other travelers, two Germans and a Jew. One of the two Germans spoke English. He is a doctor. The Jew had a large beard and long hair. The German doctor began to speak to Nawab sahib in English. At one point in their conversation, he looked at me with a laugh and said: "In India they lock bibis up like prisoners." This made me laugh too, and then he laughed even more.[16] The conversation continued through the entire meal. I was the first to excuse myself. Nawab sahib then came away too. Al-hamdulillah, the evening ended well.

Today Is the Twenty-Sixth of November

We both woke early and prayed fajr. Then, as usual, I washed up and prepared my outfit. We went down to have tea. We decided that today we would take a boat to the railway [station]. This is because the rail lines that run to Haifa etc. are far from here. The path to the railway is blocked by a lake, which can be crossed by boat.[17] They say, though, that the water level is low and that we can go by carriage as well. I said to Nawab sahib, "Let's go by carriage," and we instructed the hotel owner to have our lunch ready by eleven sharp. We are about to leave for a tour of Tiberias, and then we will take the same carriage as before to the railway. They say there is a small village there called Samaq.

We went out to visit Tiberias. On the way Nawab sahib stopped off at a Jewish man's house. The man runs a school there, which was in session. It had many students, male and female. His wife invited us to sit in their visiting room. The Jewish man soon came. He is cockeyed in one eye. Sahib asked him something about the road ahead and then prepared to leave. He stood up but the man stopped him and offered him coffee. Then we went back to the hotel. We told the owner, "We don't have time to eat before leaving. Please pack our food. We must leave soon."

While he packed the food, I wrote and mailed two letters, one each to Muhammadullah and Khurshid. Nawab sahib paid the bill. Once we were ready, we went out to the carriage. Just then, the Jewish man arrived and sat down in the front of the carriage. We set off. The road was terrible.

Solomon's [Suleiman's] Hammam

After we had been traveling for thirty minutes, I was told that "there are four hammams here that are fed by natural hot springs from the mountains." They are used to treat people afflicted by joint pain or rheumatism. People with boils also bathe here and are cured by God's command. Patients stay here for a week or two. The hammams are known as "Hazrat Solomon's hammams."[18] They say that Solomon himself used to bathe here, though only God knows the truth. Nawab sahib and I visited the hammams. They are very well-made, with clean, water-filled pools. I put a hand in to check, and it was certainly hot. The entire hammam is hot. After visiting all four hammams, we went to a nearby hill where we saw piping-hot water gushing from the earth. There was steam in the air. This water flows directly to the hammams' pools. I walked down away from the source, where the water was cooler, to wash my hands. I drank a little too because the water is so clean and pure. The taste is very brackish, as though it were thoroughly mixed with salt. Then we returned to the carriage. The Jew took his seat, and we departed.

The Train to Medina

We were still on the road when Nawab sahib said, "If you decide to do so, then we can take this railway line all the way to Medina and visit the shrine of the Lord of the Two Worlds, the Master of the Universe [the Prophet

Muhammad]. This will allow us to spend ten days in Medina. Alternatively, we can perform the hajj [in Mecca] first and then visit Medina afterward, whichever you prefer."[19]

I replied, "Wonderful. If this rail line goes straight to Medina, then, by God's command, we should certainly go to Medina first. We will enlist the support of the Master of the Universe and then proceed directly to the house of God [the Kaaba], may he be exalted. This is an excellent plan."

We agreed together that this is what we would do.

It is one in the afternoon now. We have arrived at Samaq station.[20] When we arrived, the train to Haifa was standing on the platform. When he saw it, Nawab sahib said, "Well, look at that. So that's what trains are like here."[21] We went into the station and had food in the waiting room. After the meal we inquired about the train to Medina and were told that it would leave at nine tomorrow. Inshallah.

Meeting Mrs. Suqrat

The stationmaster's name is Suqrat [Socrates], but he is a Christian and a very good man. The poor fellow approached us and said, "My memsahib would like to meet your bibi sahib."[22] Nawab sahib replied, "Of course, that is no issue." Suqrat sahib does not speak a word of English. He knows French and Arabic. Mrs. Suqrat came to meet me and took me to her home, which was nearby. The house was well decorated. I also met her mother-in-law and sister-in-law. Mrs. Suqrat speaks English, so she conversed with me in English. She said, "I only got married a month ago. I don't feel very comfortable here. When I saw you today, I immediately wanted to meet you and talk with you. I'm so glad to have met you. My mother- and sister-in-law only speak Arabic. I learned English at home, the only one in my family to do so. I am from Egypt. That is where my parents live."

Mr. Suqrat reminds me of my brother, Mirza Haidar Jivan Beg. He is very affable. He had a long conversation with Nawab sahib in French. He invited Nawab sahib up to visit the house as well and was very hospitable. He produced an angrezi apple, cut it himself, and served it to us. He then offered us coffee. I gave each of the four of them a Hyderabadi rupee. They appreciated the coins and happily accepted them. Then they arranged a room and a bed for us to stay the night in. We left at five and went to the room. I was already settled in and resting when Mrs. Suqrat came and said, "Come, I will show you around." I happily agreed and got ready.

She took me to the sea lake that we had crossed earlier in our carriage. There was very little water in the area that our carriage passed through, but a decent amount of water still managed to get inside. In this part of the lake, though, the water was quite choppy. It has a wooden bridge meant for strolling and taking in the lake air. We walked on the bridge for a while, until we were joined by her sister-in-law. The three of us walked around and chatted about this and that. Soon it was sunset and a cold wind began to blow, so we returned to the station. Here we met with her mother-in-law, who spoke to us in Arabic and taught me some of that language. Then we all went upstairs. I went to my room and told Amina bi, "Go now and make a chicken *salan* [in sauce]. Buy two chickens. One is for us; the other is for Mrs. Suqrat." Amina bi bought two chickens and made the salan. When it was Mrs. Suqrat's dinnertime, I asked for a dish, filled it with the salan, and had it returned. They all enjoyed this Indian salan. We had our meal too. Sarkar-e ala began to pray and recite vazifas.[23] I prayed isha.

Soon Mr. Suqrat came by. "Mrs. Suqrat is asking for you. She thanks you for the salan." Sarkar-e ala was busy with his prayers, so I took Amina bi with me up to their bungalow to visit her. All four of them were waiting and were happy to see me. We chatted for a little while. Then each of the three ladies sang a song. Two sang in Arabic, and one in English. They asked me to sing too, but I asked them to excuse me due to the presence of Mr. Suqrat. Instead, I said to Amina bi, "Sing them a ghazal in your best voice." She very gladly sang a ghazal in praise of the Prophet. Her singing is not the best, but it's good enough. Anyhow, they enjoyed it.

I left at nine and returned to my room. I had very bad diarrhea, the result of my having drunk water at the hot spring earlier in the day. It is very cold here now. We went to sleep, but I was awoken at twelve o'clock by an intense bout of diarrhea and a pain in my stomach. I did not tell Nawab sahib about it but went straight back to sleep. I had more diarrhea at three. Despite having diarrhea three times, I still felt healthy. After this I had no more loose motions, and I slept comfortably until morning.

November 27, 1909

We woke up in the morning, prayed fajr, and got ready quickly because the train would leave at nine. Mrs. Suqrat called me up to her home. Sarkar-e ala came as well. She gave us breakfast. The train pulled into the station while we were eating. Nawab sahib began to rush, but Mr. Suqrat said, "Don't worry. I'll hold the train." We don't have much luggage with us, just one handbag each and a goatskin for water. The rest is being stored in Haifa, where we bought our tickets. I went out to the station, said goodbye, and boarded the train with Amina bi. Nawab sahib had a chicken slaughtered just in case we need it. It was nine in the morning according to Arab time. The train does not have a bathroom. Men and women sit together in the same carriage.[24]

Now it was 1:00 p.m. The train stopped at a station in a small town called "Darrah."[25] From here we will have to change lines for Medina. That is why we disembarked at Darrah. Sarkar bought another chicken here. We gave the chickens to Amina to cook and got some warm bread from a shop. I understand that the train to Medina will pass through here at 3:00 p.m., meaning we will have to wait a full three hours. While we wait, Amina bi will prepare the chicken. Sarkar-e ala has gone to visit the bazaar. I am staying here at the station because I still feel weak after last night's diarrhea. I met an Arab coolie who helped us and is still helping us. But he did one thing that was strange. He took a bit of the chicken that was still roasting and ate it. When Amina bi tried to stop him, he said, "I am Islam."[26]

Nawab sahib returned from his visit to the bazaar with a woolen *chugha* overcoat and eight yards of thick cloth for himself. He also bought an undershirt for Amina bi for two and a half majidis because she does not have any warm clothes and it gets very cold here. Our food was ready, so we sat down in a clean space and ate. We gave the leftovers to Amina bi and the coolie. We are [now] sitting and waiting for the train to arrive.

Departure for Medina

It is 5:00 p.m. The train to Medina arrived at 5:00 rather than at 3:00, as scheduled. Al-hamdulillah, both of us climbed aboard, and the train departed. We got a little compartment, large enough for one person to lay down comfortably. I am thankful that God is taking us to this place [Medina]. This fact transforms every inconvenience into a comfort. Al-hamdulillah, we have found comfort at every stage in our journey. Now, too, by the intervention of the Prophet, things will surely work themselves out.

At every station there is a sudden flurry of Turks on the train (fig. 3.1). These Turks and fruit sellers freely climb in and out of the train wherever it stops. They all speak Turkish. They know very little Arabic and no French. The train's crew and officers are all Turks. Purdah is very well arranged for. Armed Turkish soldiers regularly patrol the train with guns and cartridges. The system is the same at every station. My compartment does not even have a door. There is another compartment directly across from mine, but it is occupied by two Turks. They are very good people. We don't have

Fig. 3.1 The caravanserai at the station in Mada'in Salih on the Hejaz Railway route, circa 1908. Photograph by J. H. Halladjian. *Source: Hajj and the Arts of Pilgrimage, Khalili Collections, ARC.al 6.40.*

any bedding with us because we left all our things in the waiting room in Haifa. We initially thought we would return to Haifa, but, by God's will and on the command of the Prophet himself, we have unexpectedly ended up heading in this direction.

The woolen overcoat is serving us well now. We have spread it out on the bench to keep us warm while we sit. We had Amina bi sit outside our compartment with our two bags. There were already many other travelers there. Our Turkish fellow travelers, who are very honorable [sharif] and kind, brought out one of their own quilts and hung it on the hooks outside our door, which does not even have a latch. They also offered us two brand-new quilts: "Please use these." God knows why they had so many quilts with them. We politely declined, but they brought out even more quilts: "Look," they said, "we have plenty." Thanks to those quilts and God's grace, we passed the night sitting there in comfort.

November 28, 1909

It was morning. The Turks themselves went and refilled our waterskins. Sarkar and I got down and went to the bathroom at one of the stations. The train stops at each station for a long time. If the train is about to leave a passenger behind, they shout out, "Stop!" and the train stops. We milled about the station for a while. We saw some fresh bread for sale. After we did our vuzu the Turks appeared with chai and bread. We thanked them and paid them. They even vacated their room, which was opposite ours, and offered it to us. They went off somewhere else. Then it was me in one room and Sarkar in the other. We were both very comfortable.

As usual, Sarkar would buy some chickens, and whenever the train stopped for long enough, Amina bi would cook them, and we would eat comfortably. At one point in the journey, we encountered two Turkish thugs who happened to notice the golden buttons on Sarkar's shirt. They kept passing by him and staring intently at those buttons. This frightened me. I immediately removed them and put them safely away. Later that night, one of those Turkish thugs pulled our bag out from near Amina bi's head and had already opened it up when she suddenly woke up and raised the alarm. By God's will, everything turned out all right. He only managed to grab one pajama from the top of the bag before running away in the moving train. That Turk never appeared in the train again. We were saved by God's kindness and the Prophet's intervention. I remained awake all night just in case he reappeared.

The twenty-ninth of November passed similarly. I slept through the night.

Today Is the Thirtieth of November 1909

This is our fourth day in the train. Today is Tuesday. They say that today we will arrive at the threshold of the door of the Master of the Universe, the Companion of God. We gathered our things. Nawab sahib took from his bag a set of Arab clothing that he had brought with him from Hyderabad, which we have been carrying with us everywhere. This type of clothing looks very smart and radiant.

Impatience for the Pilgrimage to Medina

The restlessness of my heart and the impatience of my eyes is indescribable. O God! I thank you for bringing this sinner to the Prophet's door, so I might make the pilgrimage I have so fervently wished for. Al-hamdulillah, our wish will now be fulfilled due to the intercession of the Prophet, his descendants, the children of Ali, the blessed companions of the Prophet, and the wives of the prophet. *(God! May you grant all faithful men and women and all lovers of the Prophet and his descendants a pilgrimage to the resting place of the Prophet and to those of his family. And may they perform the hajj. And may this base one, the preparer of this travelogue, who is no more than a fleck of dust on the feet of the lovers and the dear ones of God, be granted this as well. Amen.)*[27]

Notes

1. "Ashkarukum."
2. "Katira tayyab hada."
3. "'Aish kabat naha." The Arabic is unclear, and this translation is highly speculative.
4. "Ana bi-sair Makka wa Madina ba'd min naha."
5. "Ashrab."
6. "Ana ma ashrab."
7. "Enta ashrab hada sigrit Hindi."
8. Jesus is generally regarded as having been born in Bethlehem and raised in Nazareth. Begum Sarbuland may have heard a local source tell her that Jesus was born in or near Nazareth, but this seems rather to have been a mistake on her part.
9. Although Begum Sarbuland acknowledges in an earlier passage that Christians in the region speak Arabic, she steadfastly reserves the category of "Arab" exclusively for Muslims. In this and in other cases, she conflates religious and ethnic identities, seemingly unaware that Arabs could be any religion other than Muslim. This accounts for her surprise that Palestinian Christians are very good at Arabic but do not speak English.
10. This is the site where Mary's house is said to have stood and where she was visited by the angel Gabriel, who informed her that she would give birth to Jesus.
11. The Qur'an frequently refers to Jesus as the "Spirit of God."
12. The text erroneously gives November 21.
13. The sites visited are detailed in H. Khan, *Safarnama-e Madina*, 5.
14. An Ottoman coin.
15. "Hada ma baghit."
16. Begum Sarbuland expresses a similar sentiment below. She was clearly sympathetic to the man's opinion.
17. The Sea of Galilee.
18. Now known as the Tiberias Hot Baths, they are located inside Israel's Hamat National Park. The water emerges at a temperature of 60° Celsius.
19. From his own travelogue, it appears that Sarbuland Jung had already long been interested in taking the newly opened rail line straight to Medina. His interest may have stemmed partly from Hyderabad's involvement in its construction. H. Khan, *Safarnama-e Madina*.
20. This line was the Jezreel Valley Railway, itself a branch of the Hejaz Railway that ran from Haifa to Dara'a. It connected with the main Hejaz Railway line in Dara'a. The full line was opened in 1905. Samakh was essentially the final station on the Hejaz Railway in Palestine. The village there was dispersed after its capture by Zionist forces in 1948 and no longer exists.
21. A celebrated infrastructure project that showcased Ottoman Muslim modernity, the Hejaz Railway was only opened the previous year, in 1908. The track ran from Damascus to Medina. Hamidullah was fascinated by its construction and eager to see it in operation. The couple seems to have taken this roundabout tour through Palestine for just this reason.

22. The term *memsahib* generally refers to Europeans, but Begum Sarbuland clearly used it to refer to Christians generally, since she was in no doubt that Mrs. Suqrat was Egyptian. The term *bibi*, by contrast, in her usage clearly marks her as Muslim. Thus, for Begum Sarbuland, conceptions of race were intertwined with religion.

23. An honorific for her husband. The term means something like "exalted lord."

24. Hamidullah's account does not include any description of their overnight stay and interactions with the stationmaster. He records simply boarding the train and continuing the journey. In contrast with Begum Sarbuland's terse description of the journey, her husband gives his own detailed account of the route and the topographical features they saw along the way, with a particular focus on connections to biblical events. She seems focused on the train's interior; he, on what was outside. See pages 250–54. H. Khan, *Safarnama-e Medina*, 8–9.

25. Dara'a was a major junction on the Hejaz Railway and is today a small town near the Syrian-Jordanian border. She spells the name incorrectly here.

26. "Ana Islam."

27. This parenthetical has apparently been added to the book's text by either an editor or a scribe—that is, the "preparer of this travelogue"—at the press. Two more similar parentheticals appear later in the book, one of which makes an objection to a statement made by Begum Sarbuland.

CHAPTER 4

MEDINA

مدینہ منورہ

The Prophet's Servants Arrive in His Court

Ah! The land that can only be described as heaven on earth is now coming into view. May God bless the Prophet and his family.

Impassioned by our desire, we both stood up and began to look out the window, our eyes scanning frantically, looking for that radiant garden [the Prophet's tomb]. We both wept tears of joy from our impatient desire to see his tomb. Soon we could see date palms before us. Then we could see the station and the people of Medina, all of which increased our anxious anticipation further.[1]

Al-hamdulillah, the train pulled into the station at four in the afternoon, at asr. Everyone here is a Muslim and a person of the faith. The Prophet's ummah was everywhere here, wearing Medina-style abas with turbans on their heads. Ah! How beautiful and divine this clothing is. The platform was crowded with Medinans, and a few of these people were waiting for us. They came forward and, taking Nawab sahib's hand, helped him off the train. Everyone shook hands, one after the other. Nawab sahib was wearing Medinan clothing too. All these Muslims and friends congratulated us on our arrival. "May you present yourself at the fragrant tomb of the Prophet," they said. For us, this day is better than a thousand Eids. "Let's go quickly," Nawab sahib said, "so that I can pray asr in the Haram Sharif [mosque] of the tomb of the Lord of the Two Worlds."

They all had our luggage taken from the train by their own hands, and then we exited the station. We rented two carriages, one for Amina bi and me and the other for Nawab Sarbuland Jung bahadur and the Medinan nobles. Sarkar said, "I want to walk there on foot."[2]

"Ride in the carriage for now," they replied. "We will show you from where you should go on foot."

When the carriages reached the city walls, we stopped to read the appropriate prayer for entering Medina. From here we proceeded on foot. On the way, I suddenly realized that I would need to go to our accommodations to cleanse and purify myself [*ghusl*] before I could proceed to the Haram Sharif to have the honor of prostrating myself before the Chief of All Creation and attaining complete happiness in this and the next life.[3]

For this reason, I turned in the direction of our accommodations while, by God's grace, Nawab sahib continued directly to the Haram Sharif to achieve his desire to perform a pilgrimage to Muhammad Mustafa, the Messenger of God. We had obtained permission from His Radiant Highness to stay in this house before leaving from Hyderabad.[4] They were expecting me. When I arrived, the wife of the house's custodian welcomed me graciously. There were also a few other bibis who are from the same city [Hyderabad], but for the last ten or twelve years, they have been prostrate at the feet of the Messenger of Allah. They live in great happiness and contentment. May God soon also allow me to prostrate myself at the door of my beloved Prophet [*habib*] and press my eyes to his threshold so that my anxious heart may at last find peace. My Sarkar—al-hamdulillah!—has obtained the honor of presenting himself at the blessed Haram before me. May God bless him in this life and the next, and may the Master of the Two Worlds gaze upon him and bestow kindness upon him, on his children, and on his elders and through him, on me, at every moment of every day. Amen, oh, Lord of the Worlds!

I went to our apartments after meeting all the bibis. I had heard that from this house you could see the minarets of the Haram Sharif, and so, as soon as I entered, I went straight to one of the windows in this bungalow and sat down before it. I could see the blessed minarets. They were awash in light. The *mahmal* will come today, and for that reason the sultan has ordered that the minarets be fully illuminated.[5] I gazed at them with excitement until I cried tears of joy. I stayed there for long time, gazing at the blessed minarets in a state of longing and passion. I was still there when the lady of the house called out to me. "Please wash up and do vuzu. The food is ready."

I looked at my watch and saw that it was 8:30 in the evening or, according to the local time, 2:30.[6] I got up from the window and washed up. The *dastar-khwan* was already spread, but I waited for Sarkar to return from the Haram Sharif [to eat]. The food was already ready when he arrived. When he came, I went downstairs, laid out the dastar-khwan for Sarkar, and said, "By your leave, I will eat with begum sahib, the lady of the house." Sarkar happily assented. I left.

All the bibis were waiting for me. They were happy to see me. After the meal I said goodbye to the bibis and went back to our room. It is decorated entirely in the Turkish style. The sofas are arranged along the walls and are strewn with pillows. I am exhausted. I lay down on one of the couches. Sarkar did the same. How wonderful this evening is! I am filled with yearning to visit the Luminous One. I can think of no one and nothing but him, the Messenger of Allah. May God, through the Prophet's intercession, grant this pilgrimage to my mother, father, brothers, sisters, children, friends, and all my brothers and sisters in Islam. Amen, amen.

I was in a strange state all through the night, neither asleep nor awake. At midnight Nawab sahib did vuzu for the *tahajjud* prayers and then went to the Haram Sharif. I stayed at home, lost in reflection on my Lord right up until dawn.

The Land of the Beloved

December 1, 1909: I woke up and did my usual morning activities. Nawab sahib returned home. We had breakfast, and then he went back to the Haram Sharif. I went upstairs to see the bibis and mentioned to them that it would be nice if they would go out with me so that I could see the city of Medina, the very city my Prophet once walked. The city and its lanes are all blessed. They have attained the greatest honor. Two of the bibis said, "Of course we will go with you." The three of us got ready. I put on a Medina-style burqa, and we set off. We passed by the entrance to the Haram Sharif, and the moment I saw it, my heart commanded me, "Halt!" I stopped before it with reverence. I remained standing there for fifteen minutes. The effect on my heart was so intense that it cannot be described or explained. The two bibis finally brought me to my senses. "Please, let's go." At their insistence, I went.

My Lord's city is very verdant, prosperous, and populous. The food sold here is of the highest quality. Fruit sellers, grocers, and cobblers are all busy at work. There are many shops selling grains. I saw quite a bit of the city during our three-hour walk. There is always freshly cooked food to be had here. In many households here, people do not cook but simply order their meals. Every type of chutney and pickle is available. No one makes it at home. Everything you could want is available. The people of Medina are extremely polite and well-dressed. They wear clean, pressed clothes of the highest quality. Among the fruits, the pomegranates, dates, and apples are best, and plentiful too. They also have excellent olive pickle. I stopped at a cloth shop to buy some things that I will carry with me into the Haram Sharif for the children.[7]

When I returned home, the time for asr was nearly over. Nawab sahib had come back. "Are you ready now?"[8]

"I am."

"Tomorrow is Thursday. The custom is that the *muallim* will come in the morning and lead us through the Bab us-Salam, according to tradition. Once that is done, you may present yourself [at the Prophet's tomb] whenever you like. The muallim has just informed me that he cannot come now but that he will come in the morning. Our muallim has a deep and long connection with our family. He is the son of Abd ul-Karim Bartaranji."

Having heard all this, I replied, "Very well, as you wish."

Another evening passed in anticipation.

A Pilgrimage to the Tomb of the Prophet

December 2, 1909: I woke up early in the morning. The muallim arrived. I had breakfast and did vuzu. Then I changed my clothes, put on a burqa, put some rupees in my bag, and went out. Sarkar had already gone to the Haram Sharif ahead of me. The muallim and I departed for the Haram. My eyes and heart were anxiously waiting for the gateway to come into view. I have waited and wished for this [moment] for so long.

Just as I was thinking this, suddenly, I saw it. I read the fatiha and performed salaams as I passed through the gateway. Tears flowed ceaselessly from my eyes. The happiness I felt is beyond description. What a joyous moment this is, to finally present myself at the court of the Reformer of the Sinners, the Master of the Two Worlds, my beloved Prophet. I finished my salaam and stood where the Prophet used to pray. I prayed two rakats, for I have come here as a humble slave in want of succor and forgiveness. Then I prayed two more rakats to mark my entry into the Haram. After this the muallim led me toward that sacred space [the tomb], and I presented myself directly to my magnanimous Prophet. I kept my head bowed. At this moment, my heart could hold back no longer. A torrent of tears gushed from

Fig. 4.1 A view of the courtyard of the mosque at Medina, showing the Prophet's tomb and the courtyard of Fatima in 1916. *Source: Library of Congress, Prints & Photographs Division, LOT 3704, no. 71.*

my eyes as I opened myself up to him and revealed the full contents of my heart (fig. 4.1).

Before doing anything else, I joined my hands and expressed the desire of my parents to make a pilgrimage to the grave of the Prophet and to demonstrate their reverence and passion [for him]. Then I reverentially read petitions in the names of each of my children. I then shared the greetings of my brothers and sisters and their spouses and stated their desire to make this pilgrimage and show their reverence. Then I expressed the longing of my friends to make a pilgrimage to the tomb of the Prophet. I affirmed Mrs. Hydari's and Mrs. Mumtaz's desire to perform a pilgrimage to the Haram. I prayed for their faith and happiness. May God on high, through the intercession of the family of the Prophet, the descendants of the Prophet and of Ali, and the Prophet, grant the prayers and wishes of this sinful one. Amen, amen! O Lord of the Worlds! Honor this lowly one who longs for you. Fill my heart with the heavenly light of God. Amen.

I sat there, deep in longing, in a state of intense passion. I cried ceaselessly as I opened myself up to His Eminence. I explained my piteous state directly to the merciful Prophet Muhammad Mustafa ibn Abdullah bin Abd ul-Mutallib bin Hashim. My heart refused to let me leave. After a very long time, the muallim brought me to my feet saying that there were more tombs and places that I must visit.[9]

The Graves of Hazrat Abu Bakr and Hazrat Umar

I rose to my feet. We walked a short distance, and I then presented myself at the graves of Abu Bakr Siddiq[10] and Umar.[11] I gave my salaam. The graves of these companions[12] are near the Prophet's. Salaams are made here after walking just a few steps.

In the Court of Hazrat Fatima uz-Zahra

A few more steps brought me to the tomb of the daughter of the Prophet, the first lady of heaven, the best among women, Fatima uz-Zahra.[13] *(Of Fatima uz-Zahra, the Lord of All Creation stated in a famous hadith, "Whoever has oppressed her has oppressed me. And whoever has oppressed me has oppressed God. And whoever has oppressed God, no matter what good deeds they have done or will do, every justice-minded Muslim may deduce how they will be treated on the Day of Judgment." Another hadith says, "Whoever harms Fatima harms me." How can anyone who has harmed the Prophet expect him to intervene on their behalf on the Day of Judgment? May God protect every faithful Muslim, male and female, from the anger of Fatima, and let them not antagonize her.)*[14] I felt a wondrous sensation while presenting myself at her grave. I stayed there for a long time and made a heartfelt, passionate salaam to her. I beseeched her. "Bibi! Please speak kind words to your father about this base woman." The tomb of the Lady of the Plain of the Day of Judgment is located immediately beside that of His Exaltedness. After performing my salaam, my heart patently refused to leave this place because of the bounteous favor that came from focusing my eyes on the grave of the Lady of Judgment Day. At long last, though, I turned again to face the Prophet's tomb and prayed a fatiha.

At the Blessed Tomb of My Lord, Amir Hamza

I turned toward the holy tomb of Amir Hamza and read a fatiha.[15] I then presented myself at Gabriel's Door, through which, I have heard, Gabriel would pass when he delivered divine revelations and celestial commands to the Prophet. I prayed a salaam for the four archangels and for all the other angels.[16] Then I went to the women's area of the Haram, where all the women sit to pray *duruds* and vazifas after completing their pilgrimage and performing their salaams. I sat there too and began to pray duruds. I stayed there so long that, by the Prophet's grace, I was able to perform the zuhr, asr, maghrib, and isha prayers there.[17] Al Hamdulillah, Nawab sahib and I were blessed to perform these prayers with the congregation. We both returned home at 9:00 p.m., ate, and fell asleep.

A Great Assembly of Lovers of the Prophet

We woke up at one in the morning, did vuzu, and went to the Haram to perform tahajjud prayers. Here we saw that the Prophet's shrine was already filled with people from his community who had also come to pray tahajjud. Seeing this great gathering of the Children of Unity brought a distinct pleasure to my heart. Allah! Allah! These passionate pilgrims do not just attend the five daily prayers! Countless men and women also come to his threshold to perform the tahajjud prayers. We made our salaams and prayed tahajjud and then began reciting duruds and vazifas. We continued until fajr, which we prayed according to the Shafii reckoning.[18] The benefit of performing fajr in the Shafii manner is that instead of completing two prostrations for each *surah*, they read Surah Sajdah and thus perform three prostrations per surah.[19] After the prayer we continued our duruds and vafizas until sunrise. When I finally lifted my eyes and looked around me, I saw something wonderous. I was surrounded by a sea of men and women all lost in meditation, invoking the Divine.

Departure from the Blessed Threshold

December 3, 1909: In the morning, after finishing our vazifas and the *ishraq* prayer,[20] we salaamed every grave before returning home. There we renewed our vuzu and had breakfast. Today is Friday. I heard that tomorrow a caravan will leave for Mecca because the hajj is near. Some Bedouins have taken to creating a nuisance for the caravans, so they will leave three days earlier than intended. This means that we will also need to leave soon, whether we travel by camel or by train. This news upset me. My heart was perturbed, so much that I cannot describe it. I went into a stupor for several minutes. Ultimately, though, I bowed my head and acquiesced to the Lord of Medina's command. Nevertheless, it seems to me that, inshallah, by the grace of God and his Prophet, I will return here again. The thought brought me some measure of peace, but my restlessness did not disappear and my heart was in constant anguish. We spent half the day in the Haram Sharif.

It is a wondrous thing that when I am near his tomb, my restless heart is completely at peace. Aside from my focus on God and his Prophet, no other thought can enter my heart.

We prayed zuhr in the Haram Sharif and then went to visit the Jannat ul-Baqi cemetery.[21] The journey took thirty minutes. We finished visiting all the necessary graves by asr. I wrote my name and the names of my parents on a small board. I also wrote the name of one of my close friends, Ummat ul-Islam Begum, who belongs to the family of my pir. She had asked me to write her name on a board and to bury it in the Jannat ul-Baqi. Then we left.

Nawab sahib had arranged for a reading of the Maulud Sharif in the Haram Sharif after the maghrib prayer, so we went straight there.[22] Everything was ready. The Maulud Sharif began after the maghrib prayer. It lasted for a full hour, and many people participated. When it was finished, everyone prayed isha. After isha the muallim distributed sweets. Allahu akbar! By the grace of his Prophet, these sweets were so imbued with divine blessing that, with no exaggeration, they were distributed to every single person in the Haram Sharif. Everyone was so excited to receive it that they sought it out themselves. They were delicious. Maulana Maulvi Abd ul-Baqi, the son-in-law of my pir, Maulana al-Hajj Maulvi Muhammad Abd ul-Wahhab, had them prepared under his direct supervision. Any item prepared by such a venerable person that has been accepted by the Prophet in his great kindness and love for his ummah is sure to be excellent.

Once we had finished, as is proper, we read a salaam at every grave before returning home. Then we ate and went to bed. As usual, we arose at one in the morning and went to the Haram Sharif. My heart found a bit of solace here. After praying tahajjud we read vazifas until fajr. We also prayed ishraq and made salaams at the Prophet's grave.

Today Is the Fourth of December

Saturday morning, after fajr and ishraq, we went for salaams as usual. It is impossible to describe my frenzied state. It feels as though my heart will burst out of my chest. God! The Prophet's love and attraction are so vast. We sinners are nothing compared to it. My heart is so drawn to and so in love with the Prophet that it feels as though it is being ripped from my chest and left to writhe madly in its sorrow from separation from him. Only those who have experienced it can understand the sensation of having to depart from the Prophet's grave. The pen lacks the power to record it. The tongue is powerless to declare it. It felt as though my soul might depart my body to remain in this blessed place. Our hearts, Nawab sahib's and mine, commanded us to recite salaams over and over. We did, repeatedly, but it never seemed to be enough because it never could be enough. Our hearts screamed, "How can I leave such a Master? The Master who helps and supports sinners like us. It is through him that we are alive. He looks after us at every moment of our lives." We prayed: "O Prophet of God! O Son of Abdullah! O Hashimi and Matlabi! O Master of Mecca and Medina! Having visited you, we now wish to perform the hajj and to visit the Kaaba. Please attend to us. Ask God to accept our hajj in yours and your family's name. Ask him to forgive us and ours. Bestow your gentle gaze on all Muslims."

We finished our supplication and left the Haram Sharif with heavy hearts and tear-filled eyes. As we departed we turned back to look upon it again and again. Alas! How quickly we have had to depart from this fragrant garden. Yes, my restless heart! Our eyes, which may only gaze on the physical world, have become separated from it, but our hearts' eyes continue to gaze directly upon it at every moment.

We were lost in these thoughts until we reached our accommodations. Our things were ready. I went upstairs to meet with the wife of the owner of the house and the other bibis. I gave them an offering appropriate to my status. The bibis accepted it gratefully, gave thanks, and prayed for my wellbeing. Then I went back downstairs and went outside. Abd ul-Baqi sahib, my pir's son-in-law, was waiting. I gave him an offering. He elevated my status by accepting it. He then kindly accompanied me to the station. I protested that he should not inconvenience himself, but he insisted.

On the way we encountered Abd us-Samad Kashmiri. He is a servant of the Master of the Universe and a very godly man. He is a spiritual sibling to my aunt Naushaba Begum. My deceased aunt had once had a house constructed here; he is looking after it now. He asked me to come and see it. "It is not finished yet. I have already invested five hundred rupees in it myself." I accepted the invitation and went to see it. It looks as though it is about 2,500–3000 *lokets* [in size]. I also met his wife and children. I gave him an appropriate offering, which he was reticent to accept.

He also accompanied us to the station. Along the way he told me stories about my deceased aunt that I listened to attentively. He wants to have a *vaqf* stone affixed to the house in her name so that she may accrue a continuous supply of religious merit.[23] "Very well," I replied, "I will discuss this with Father and follow up with you in writing. You may proceed according to his recommendations." We were still discussing the matter when we arrived at the station. Today a massive camel caravan will leave for Mecca. I'd estimate that it contained about a thousand camels. The *shaghdaf* is very strange; just looking at it makes me feel afraid. When I first saw one, I made a prayer: "God, by the grace of the Prophet, please let there be camel carriages rather than these shaghdafs." Thankfully, by the Prophet's kindness, we ended up traveling by train.[24] The train was already on the platform when we arrived. I salaamed everyone. When I stepped close to Abd ul-Baqi to salaam him, he moved a bit closer and read a prayer for me in his blessed voice. He said, "Be at peace. You will return here, with all your honor intact, to visit the threshold of the Prophet again." Coming from such a great man, these words brought me a measure of peace.[25]

Going to Mecca

I gave a salaam and boarded the train. Sarkar said his goodbyes and boarded too. Amina bi boarded, and our baggage was loaded. This train is much more comfortable than the last. There are two large compartments inside [each wagon]. Each has two benches on either side. If you wish, you can shut the doors and lock them from the inside. I gave one of the compartments to Amina bi, and we occupied the other ourselves.

The train pulled away, and, sadly, the land of Medina slipped from view, though even now in my heart I am standing in the Haram Sharif. I am at the most perfect grave of the Prophet. My eyes remain filled with the vision of these holy places. My filthy body was departing Medina, but my soul is still in his holy court with my palms pressed together in supplication. My deepest, truest wish is that my base, physical body should remain there.

By the Prophet's grace, we did not encounter any difficulty or discomfort during the journey for, this time, I secretly hid some coal in a small brazier and entrusted it to Amina bi so that we could cook food for our meals. I also packed some ghee, salt, pepper, and [a meal of] *khichri alu*. I also have a bit of sweets from the Maulud Sharif and some cooked food too. Our entire carriage is empty. As usual, at every station Turks and Arabs nonchalantly climb into and out of the train. By chance, after two stations we once again met two Turkish soldiers who helped us throughout the journey. They even gave us some *tabarruk* from Medina, and we gave them some of the Maulud Sharif sweets. These Turks wrote their address down and gave it to Nawab sahib. These Turks speak a little Arabic; otherwise, their primary language is Turkish. We passed the night in peace and safety.

December 5, 1909: We spent the day watching the stations pass by until evening. The train has electric lighting, but it doesn't work. I heard that they disabled it for fear the Bedouins would use the light as a target and fire upon the train. We slept comfortably in the darkness.

Fig. 4.2 The Hejaz Railway line cuts through the mountains near the city of Tabuk, 1908. *Photograph by J. H. Halladjian. Hajj and the Arts of Pilgrimage, Khalili Collections, ARC. al 6.23.*

December 6, 1909: We woke up in the morning and, as usual, prayed and read vazifas. This train, we learned, has a bathroom, but it is filthy and smells terrible. The tap has no water. It's a pity that the trains in this region have no proper sanitation. I hope this will be addressed in the future. "First class" and "second class" are just names. The fact of the matter is that every class is the same, except perhaps that there are a few cushions in first class. There is no third class. The fare is as much as twenty or twenty-five rupees. The train twists back and forth as it winds its way through the mountains. We passed several towering mountain peaks (fig. 4.2). When it was evening, we slept.

December 7, 1909: Today is the fourth day since we left Medina. We both woke up at dawn. Nawab sahib told me that we would arrive in Damascus at 3:00 p.m.

Notes

1. Hamidullah's account of arrival in the city does not reveal similar emotions but instead emphasizes that he had been there before: "As one approaches the city . . . a sort of spiritual effect begins to distinguish itself. Those who have been to Medina previously point out the [city's] various blessed sites to the new arrivals." H. Khan, *Safarnama-e Madina*, 13.

2. As an expression of humility and piety.

3. She remembers that she is still on her period and temporarily unable to visit the sanctuary. There are still two days left on her period. Once those passed, she would have taken a full bath (*ghusl*) to cleanse herself before praying or visiting the Haram.

4. A reference to the nizam, who owned properties in the holy lands, including particularly charitable endowments for pilgrims. This house may have been reserved for the use of dignitaries. Hamidullah refers to it as "the rabat in Shamsiyyah Bagh" that had belonged to the family of the deceased Hyderabadi prime minister Asman Jah (1839–1898). H. Khan, *Safarnama-e Madina*, 13.

5. The *mahmal* was a palanquin sent in a huge procession by the sultan near pilgrimage time to symbolize his protection of the holy lands. The primary mahmal came from Egypt, but they could also come from other regions too. That night, the lights commemorated the arrival of the mahmal from Damascus, a point recorded in her husband's journal.

6. The Islamic day begins at sunset. Thus, this line may mean that it had been two and a half hours since sunset and was thus 2:30.

7. With the intention of then giving her children clothes that were blessed by their presence in the Haram Sharif.

8. He asks if her period has finished.

9. This passage underlines the personal and private nature of her diary. Most published accounts of the Prophet's tomb include a detailed description of the space and only general notes on the prayers they performed, but here she does not describe the tomb at all.

10. A companion of the Prophet and later the first caliph of Islam after the Prophet's death.

11. A companion of the Prophet and the second caliph.

12. The Prophet's disciples and followers during his lifetime are known as "companions."

13. The daughter of Muhammad and his wife Khadija. The location of her final resting place is uncertain, with this site visited by Begum Sarbuland being one of two likely candidates.

14. This parenthetical was probably added by the book's editor or scribe. Less likely, it might have been inserted by Begum Sarbuland at a later date. The language and content here is significantly at odds with the rest of the diary's entries, which are devotional rather than scholastic. These parenthetical insertions also appear to have been made by a devout Shia, while Begum Sarbuland was confessionally a Sunni.

15. Hamza ibn Abd ul-Muttalib was a prominent early Muslim military commander. His military exploits are legendary and have become the subject of vast epic romances that were once extremely popular in India.

16. In Islam, the four "great angels" are Jibril (Gabriel), Mika'il (Michael), Israfil (Raphael), and Azra'il (Azrael).

17. That is, from noon until night.

18. The legal schools of Islam differ in minor details regarding Islamic practice. Individuals generally follow one school, but moving between the schools is permissible. Here Begum Sarbuland prays fajr according to the Shafii school rather than her normal Hanafi school.

19. The fajr prayer stipulates the performance of four prostrations divided into two sets (rakats) done in conjunction with a reading of two individually selected surahs. However, the recitation of Surah Sajdah uniquely requires three prostrations. Thus, by selecting this surah, one may permissibly prostrate six times in two rakats.

20. A supererogatory prayer that is offered after sunrise.

21. A major cemetery in Medina, it contains the graves of many important figures from Islamic history.

22. A recitation in honor of the Prophet.

23. That is, he would like to turn the house into a charitable foundation and have the religious merit accrued to the deceased.

24. Rather than go directly to Mecca by camel caravan, they instead take trains to Damascus and Beirut before sailing south toward Jeddah where she finally must travel in a shaghdaf.

25. She did return, in 1934 and in 1937.

CHAPTER 5

THE LEVANT AND SINAI

شام و جزیرہ نما سینا

A Visit to Damascus

I told Amina bi to prepare our belongings, for we would be getting off the train today. Those two Turks are still with us; they've been looking after our luggage. The train arrived in Damascus just after three. We have encountered many large stations on this journey, each of them having good sanitary arrangements and water taps installed at regular intervals. The station in Damascus was also big and bustling. A Cook representative was waiting for us when we disembarked as usual, but we also met someone connected with an Arab hotel. He praised his hotel extensively and suggested that we come and stay in it. "You are Muslims, and so are we. You must certainly come to our hotel." Responding to his insistence, Nawab sahib asked the man for the hotel's name. And what a sweet name it was! "The name of my hotel is Medina Munavvara, 'The Radiant Medina.'" I immediately said to Nawab sahib that we must stay in this hotel; I was sure to like a hotel with such a name. Nawab sahib was pleased too. The owner of the hotel picked up our luggage and placed it in a fine carriage. Arriving at the hotel, we learned that the daily rate was ten annas per person. We were surprised that the hotel was so cheap. Up until then, we had only stayed in German or English hotels, whose rates were around eight to ten rupees per night. This is the first time we had encountered a Muslim hotel.

We went inside. The stairway was made of marble. Each step was carpeted. We were pleased that the hotel appeared to be so nice. Upstairs, we were given a room containing six beds, but all of them were soiled, and there were cobwebs in the corners. Though the building itself was exceedingly grand and finely built, the furnishings were anything but. The table was filthy, and the mirror was cracked. I had noticed that in English hotels they would put a sheet beneath the blanket so it would be kept free of sweat. After a traveler had used the sheet, it would immediately be exchanged for another one, and in this way the blanket would always remain clean and devoid of odors. Here, in place of blankets they had quilts—quilts so filthy that they reeked of perspiration. Oh, when will these people learn proper cleanliness? They don't even put a sheet beneath their quilts! The room is very beautiful, with excellent floors made of marble and black stone. It is a pity that the rooms are not kept clean.

Typically, in an English hotel there is a constant supply of hot and cold water, but there was no hot water here. There was only one dirty, smelly towel. I said to Sarkar, "Well, now that I'm already here, I'll stay the night, but only because the name of the hotel is blessed and because it is owned by people of our faith. Otherwise, I wouldn't stay here for even a minute." We arranged for an additional room and told Amina bi to sleep there. That's when I noticed how foul the bathroom was. It had such a stench that it made my brain rot. Inside there was a squat toilet. There was water everywhere. In the bathroom there was a dirty tin cup and a bucket of water. The bucket had a spout. The bathroom floor was also made of marble but was filthy. I declared that I could not bring myself to use that toilet. Nawab sahib sent for a "piss pot" to be purchased for me. The bathroom was so small that it was difficult to close the door. The owner told us that the rate of ten annas per night only included our stay—food, tea, and coffee were not available.

Once I had seen all of this, I said to Sarkar that it was only four in the afternoon, but the bathroom was not clean, and I wanted to take a bath before we went out into the city. I've heard, I told him, that the lustrous head of the Lord of Martyrs, the Oppressed of Karbala, the One Slain by Oppression and Cruelty, the grandson of the Prophet Muhammad-e Mustafa, Imam Husain—upon whom be peace—is buried here in Damascus.[1] There are also tombs here belonging to other prophets and saints. When I learned this, I was elated that we would have the honor of visiting the shrines and tombs of so many great figures. Nawab sahib replied, "All right, I will ask the man where the bathing place is."

Sarkar then inquired with the owner, who said: "There is no hammam in the hotel, but there is one in the city."[2] When I heard this, I said that it was very inconvenient for the hammam not to be in-house but in the city.

"How am I to quickly bathe and be ready to go out now? I'm still dirty from the train, it's time for the afternoon prayer, and we are only here for one night!" Sarkar told me, "The way things work here is that men's and women's hammams are in the city. Everyone, male and female, goes there to bathe. It doesn't matter whether you're rich or poor; everyone goes. The hammams are relaxing and very nice, and there are always people bathing in them." I replied that if that was the practice, then fine, take me to the hammam. "I'll have a quick bath and then go and visit the shrines." I put a pair of clothes, some soap and flour,[3] a comb, and so forth into a small bag and left Amina bi at the hotel to keep an eye on the luggage. Sarkar got out some clothes too, saying that he would bathe as well.

"Everyone's Naked in the Bathhouse"

We set off and found two hammams nearby: one for men, the other for women. I knocked on the door, and a woman appeared. When she saw me, she said, "Welcome, come in, come in."[4] I went inside. Sarkar told me to bathe quickly: "I'll have a quick bath and then come out and wait for you." I agreed and went inside. I saw thirty or forty women gathered there. There were many small rooms in which rugs and carpets had been spread. In the very center of the hammam's courtyard was a ceiling of translucent glass, such that sunlight could enter, but the wind would stay out. In the middle of the courtyard was a pool filled with water. Light-skinned, rosy women who had finished with their baths were busy getting dressed. Others were taking off their clothes and heading into the hammam to bathe. All the men and women in Damascus are very light-skinned, with the exception of the Africans.

I paused, thinking to myself that this was quite a dilemma. I'll wait to begin my bath, I thought, until everyone else has finished. This being my intention, I was standing there waiting when a woman came up to me, took the bag from my hand, and began to remove my clothes. I said: "Wait, wait!"[5] She retorted, "No, hurry!"[6] and said God knows what else in Arabic. I acquiesced when I saw that she had no intention of listening to me. She removed my clothes. I wrapped myself up in a towel, and she took me by the hand and led me away. When I entered the next room, *oho!* It was so hot my feet were immediately burned. There were vents in the ceiling above that allowed natural light to filter in and the hot steam to escape. I said, "Hey, you tyrant! Hang on, my feet are burning!" But she kept on dragging me, saying, "Come on, come inside." We passed through three different levels of the hammam and entered the fourth, where I saw many women seated barefooted before water taps, busy washing themselves. Hot water comes out of one tap, and cold water from the other. There are small basins installed near the taps that fill with water when you release them.

The woman began to rub my body, washing me. In fact, three or four of the women stopped bathing themselves, came over, and began to scrub me as well. In Arabic, they asked me, "Who are you? Where are you from?" I said to them, "I am Indian. In India, the hammam is at home. In town, there is no hammam, but here, I saw the hammam is in the town; in the home there is none. I am astonished."[7] With great effort, I managed to answer a few of their questions. These women washed me more thoroughly than I'd ever been washed in my entire life. Here I was in a huge rush to leave, and they kept on pouring out more and more water!

At last, I managed to escape them. I wrapped the same wet towel around myself, so I would not be entirely naked when I exited the hammam, but all of them called out to me, saying, "Hey, sister, [only] women here."[8] I said, "Yes, but I want a towel."[9] I began to leave, but they took hold of me, saying, "Don't go, don't go, sit, sit down for just a while more."[10] I replied, "I don't wait. I am in a hurry, husband at door."[11] They reluctantly released me. I exited the bathing area. When I first entered the hammam, my feet were burned from the heat; now, on exiting, the very same floor felt cool to the touch. I rushed to the exit and took a seat on a bench in the courtyard. A woman came and began to dry my hair. She asked me, "Oh, Indian, are you well?"[12] I replied, "Very well."[13] To make a long story short, I got dressed and handed the woman a shilling. I said goodbye, picked up my bag, and went out to find Sarkar waiting for me outside. We returned to the hotel to give Amina bi my bag and immediately set off.[14]

The Umayyad Mosque

We went to the famous Umayyad Mosque first. We went in a rush so that we would not miss the maghrib prayer. We passed through the Damascus bazaar on the way. The bazaars here are roofed, so the people inside them remain protected from the elements. Typically, bazaars are open-air. They often have canals of flowing water. All the people of Damascus have white skin and a rosy complexion. The women's dress is entirely "European," and even their hair is kept in the same style as that of "European ladies," though they do not wear hats. Whether rich or poor, whenever ladies here go out for a walk or go shopping, they put on a simple skirt. On their heads they wear a shawl of the same color. This shawl extends down to their hands. Before leaving home, they draw a niqab over their faces, put gloves on their hands, and pick up an umbrella. Oftentimes, older women will tie a handkerchief around their heads, but their outfit is always the same, either something resembling a "dressing gown" or else a skirt and blouse akin to those European ladies wear. Children wear a similar frock, and many of them tie a handkerchief over their face as high as their ears. Here, the custom is for all Muslim women to go out in the city without any big to-do. This is a great convenience for them—may God grant the women of India the same freedom from their unjust imprisonment. There was another thing here that pleased me greatly—namely, that whenever women go out in the city, the men keep their distance and treat the women with respect. This is in contrast with India, where women have no respect before men. Oh, how unfortunate are the people of India!

Two-horse carriages are ubiquitous and very cheap. Phaeton carriages and other vehicles are also widely available. In our India, the common people use simple *yakkas* and *tangas* [horse carts], while here fine carriages are the norm. We all finally arrived at the doors of the mosque, taking in all the sights along the way. When we arrived, there were still ten minutes until maghrib, so I began to examine the mosque from top to bottom. I am incapable of giving a fitting description of it. It is extremely large and very beautiful. The floors are covered in exquisite carpets. Lighting is provided by ornate glass lamps, and the walls are all built of marble. The courtyard of the mosque comprises a large open-air floor, the whole of it made of marble. At that moment, all the mosque's chandeliers were illuminated. All the wall lamps were lit as well.

Offering Prayers in the Umayyad Mosque

The leader took his place on the pulpit. The call to prayer began, recited in the most melodious tones. We first performed two rakats upon entering the mosque. The *takbir* reached our ears, and all those who had come for the prayer got to their feet. It seems that here the takbir is given by several callers posted throughout the mosque. Because of this arrangement, you can easily hear the takbir at the appointed moment for bowing and prostration no matter how near or far you are from the front. Although I was in the back, I heard it clearly. Al-hamdulillah, I prayed maghrib. When I began to have a look around after finishing my prayers, what did I see but four different teachers, each sitting in his own corner—one for the Hanafis, one for the Shafiis, one for the Malikis, and one for the Hanbalis—giving sermons and guidance to the people.[15] Masha Allah, how wonderful it was! The splendor of Islam has spread widely here! Every task is performed entirely in keeping with the dictates of the noble sharia. I stayed there for a while listening to the sermons and lessons. Unfortunately, these were given in Arabic and Turkish, and I could not understand them.

At the Tomb of Yahya

We then visited the holy shrine of our lord Hazrat Yahya.[16] This blessed tomb is located inside the main hall of the mosque itself. It is surrounded on all four sides by an ornate railing made of glittering silver and gold. The grave is elevated and very large. We lifted the covering and, standing at the entrance to the tomb, prayed a fatiha and made a salaam.

The Tomb of Hud[17]

After this, we went to the shrine of the prophet Hud. The location is only indicated by some green stones set into the wall. I have heard that his blessed tomb is below these stones. We prayed another fatiha and salaamed the saint.

The Blessed Tomb of Hazrat Khizr[18]

Having received blessings in these places, we then presented ourselves at the shrine of Hazrat Khizr. Here too were those same markers set into the wall. We both made salaams in the name of Hazrat Khizr.

The Blessed Shrine of the Head of the Martyr of Karbala, Imam Husain

After having seen everything here and, al-hamdulillah, having been blessed by our visits to these shrines, we went to present ourselves before the Martyr of Tyranny and Injustice, the forbearing youngest grandson of the merciful Prophet, our dear Imam Husain. I had heard that the head of my dear imam is buried here. We presented ourselves and stood reverentially before the shrine. The moment we arrived, my heart was overwhelmed by a strange sensation that I am incapable of describing. Nawab sahib said, "Just look at the effect that this place has on our hearts." "Yes," I replied, as we both began to weep. I wished I could stay there forever. A full hour passed, and yet we had not moved. Our hearts were unwilling to leave. It was only when the call to isha prayer sounded that our hearts came to their senses. We again recited a fatiha and, giving our respects once more, took our leave. Allah, Allah! Only those who have experienced the kindness and mercy the family of the Prophet bestows on the members of this sinful community of believers may truly know the joy it brings. I am unable to describe the sense of tranquility and comfort that descended upon my heart here and the extent to which I was affected by this experience.

We then went into [the mosque] and joined the group gathered there for prayers. We were blessed to offer the isha prayer here, praise God! People filtered out of the mosque, and we too exited and leisurely made our way back to the hotel, visiting the bazaar on the way. The hotel staff had purchased food and kept it ready for us. We ate and then had coffee. The coffee here is extremely flavorful. We drank it and went to bed.

December 8, 1909

In the morning we woke up and prayed fajr. Al-hamdulillah. After breakfast, I very much wanted to go back to the hammam to bathe again, in part because it gave relief to my body but also because it was one of the more memorable experiences of our trip. I expressed these thoughts to Sarkar, who asked the hotel owner if there were any hammams open in the morning, for if so, we should like to go. He replied, "No, not open until after noon."[19] As we were planning to catch a twelve o'clock train, I said that it didn't matter, leave it be. But then, a short while later, the hotel manager came and said, "There is a hammam opening in thirty minutes. My wife is coming now. She will take your wife there and show it to her." I got my bag and put my clothes and things in it. Once I was ready, I sat to wait so that we could depart the moment the Arab woman arrived. Soon an Arab woman wearing a beautiful Turkish skirt and a niqab appeared before me and said, "Bismillah, please, let's go."[20]

We went out with her. Sarkar went to wait for the train. This woman took me to her house. I particularly liked it because it was much cleaner than the hotel. A canal of water ran through it. The house was two-storied. She showed me the first floor before we proceeded to the second. Here I found a large room with the floor covered by a beautiful carpet. There were beautiful sofas and chairs. The hotel owner's mother was seated on a soft mattress set on a raised platform. The poor woman was very weak. When she saw me enter the room, she raised herself up with the greatest difficulty, saying, "Please, please, come in."[21] I took this bibi by the hand and helped her to sit down again, saying that she should not trouble herself. Then the hotel owner's sister and young daughter appeared. All these women were dressed in European clothing. The child was wearing a frock. They were all very beautiful; their complexion was extremely fair.

I don't like the style of Damascene houses. For one thing, they are all several stories high, and the rooms are often dark. This room, however, was very bright. Soft, cushioned seats lined the walls for people to sit on. I went and sat near to the hotel owner's mother. All these bibis spoke Arabic. Unfortunately, I could only understand them with difficulty. They were very hospitable toward me. They served me coffee. Coffee here is drunk without milk, yet it is still very delicious. It is consumed from tiny little cups called *finjan*.

Twenty minutes later, the hotel owner's wife got up, put on her niqab, and said, "Come, sister, we must go to the hammam."[22] "OK," I replied.[23] I said goodbye, kissed their hands, and left. The custom here is to join hands and kiss one another at the time of departure. The bibi and I arrived at the hammam about ten minutes later. Just as before, I went inside. I had a relaxed bath, got dressed, and then left again. I gave a *beşlik*[24] to the hammam attendants and eight annas to the hotel owner's wife. This made her very happy.

I met Sarkar while we were still on the way. We returned to the hotel to collect Amina bi and the luggage. We paid the hotel owner what we owed him. The hotel owner's wife came to me once more and said, "Come, I will show you around." I said to Sarkar that we still had a little time and that, if he so willed, I would go with this woman for a while. Sarkar assented, and I left with this bibi. She took me to a shop, the owner of which appeared to be European. I asked him, "You speak English?" He said, "Yes." I bought a warm pair of gloves and three beautiful frocks that caught my eye and quickly returned to the hotel.

Going to Beirut

Sarkar was waiting for me when I got back to the hotel. We went outside, where we found a beautiful phaeton waiting. We climbed in and headed for the station. The train was already on the platform when we arrived, so we easily boarded and took our seats. We had Amina bi sit elsewhere. The carriage Sarkar and I were in was extremely comfortable. There were velvet cushions and nice, clean benches to sit on. Strangely enough, although we were in first class, there was no bathroom. We were told the journey will take fifteen hours. We are to disembark at Beirut, inshallah.

Al-hamdulillah, the train departed and traveled along a canal for miles and miles. This canal runs through Damascus too. Here and there we saw gardens, trees, and little settlements. We saw huge mountains rising high above us. All in all, it made for a wonderful tour. Then it was evening. Sadly, there was no moonlight, so we could no longer enjoy the views. They say we will arrive at Beirut at eleven tonight.

Beirut is getting closer. A few stations before Beirut, two gentlemen got into our carriage. I lay down to rest. Sarkar continued to sit. One of them left the train again, and one remained seated. He began to talk to Sarkar. In his hand he held several bundles of narcissus flowers. He gave a few of the bundles to Sarkar before getting off at the next station. We were very comfortable there. Then, while the train was still moving, the guard came by to collect our tickets. The system here is that the tickets are collected from the passengers one station before arrival. This meant that we were just one station away. Two hours later we were in Beirut. The station is in town.

In Beirut

The train stopped. We both got off and retrieved our luggage. Here the coolies all swarm around you and eat you alive. One says, "I'll carry your things," while another goes on running his mouth simultaneously. We selected two coolies with great difficulty. There are many hotels here, all of them nearby. It was already midnight, and I wanted nothing more than to go to the hotel and sleep. As is standard here, the train was met by representatives from various hotels who came and asked us to go to their hotel. "Come with us, come with us!" Each of them sang their own hotel's praises. They are all so annoying. Sarkar told them that we were looking for the Cook representative, but they just wouldn't listen. I said that we were only staying here for a single night and that the night was already half over anyhow. "Come, let's just pick one of their hotels and go there." Sarkar agreed, and we selected the closest one. It was called Kaukab us-Sabah. This one also belongs to a Muslim. It had the same grandeur as the one in Damascus, with marble and everything, but was still filthy nonetheless. Well, at least we made it through the night. Al-hamdulillah.

December 9, 1909: I woke up and did my usual morning routine. Then I prepared my outfit and got dressed. Nawab sahib had tea and then went to make arrangements for the ship. After he left, I put on a burqa and went outside. I walked around for a while and had a look at the buildings. I didn't go too far for fear of losing my way. There is a tramway here. I was still looking around when Sarkar came and said that he wanted to take Amina bi and our luggage to the ship. "Everything has been arranged. You please wait in the hotel for now."

"Very well," I replied.

Sarkar took Amina bi and the luggage to the ship while I returned to the hotel and practiced my Arabic with the owner. I'm now able to speak just enough to cover my needs. Later a pearl seller came by and began to show me his wares. I picked out a few that I liked, but they were overpriced. I didn't like that and returned the pearls to him. I went back to practicing my Arabic with the hotel owner. Sarkar-e ala soon came and said, "Let's go. I want to show you Beirut. We'll sit in the tram. The ship won't leave for a while." "With pleasure," I replied, "I'm ready to tour Beirut." We said goodbye to the people at the hotel and left. We walked a short distance to a tram stop and climbed aboard. We spent two hours exploring the city on the trams. On the tram I met several mems. I also met a Jew, but everyone I met spoke only French or Arabic. I only met one lady, an American, who could speak English. I spoke to this American lady for a while. We exchanged addresses.

At Sea, Again

We got off the tram after two hours and went straight to the port. We took a small boat out to the ship. This is a German ship too.[25] God only knows why German ships are so filthy. I hate this German company's ships!

This ship is larger than the last German ship but just as dirty. The cargo even includes horses, cows, and chickens. These animals make everything stink. I went straight to my cabin. The cabins have four berths each. They are all right. I had just arrived when four mems came in. Three of them stayed, and the fourth left. The three mems and I each arranged our things and got settled in. We kept glancing at one another, wondering who was going to break the ice. It ended up being me. I asked, "Where are you coming from? And where are you going?" I asked these questions in English. None of the three answered me. They just looked at me and laughed. From the way they were laughing, I could tell that they had not understood my language. I said a word in Arabic, and one of them smiled and said, "I know Arabic. Don't know English. Don't know French."[26] We chatted, and I learned that the second woman only knew French, and the third, Russian. We all had a good laugh at the situation. There are four ladies all in one place, each of them from another country, and none of them knows the language of the other. All they can do is stare at one another. Of course, I can speak a bit with the lady who knows Arabic. If we also use hand gestures, we're able to understand each other a little. The Russian and the French lady, though, can do nothing but sit and laugh. All three of them are wearing English-style clothing. They are uncouth and entirely savage in their manners and etiquette. They are ungracious and have foul moods, except for the Arab one. She is all right.

The dinner bugle sounded at eight. We all went to eat. After eating I retired to the cabin and slept.

In Jaffa

December 10, 1909: Our ship arrived in Jaffa in the morning. I woke up, washed up, and got dressed. Nawab sahib came to me. "Let's eat breakfast and have a look around. The ship has already arrived in Jaffa."

"All right," I replied.

We went to the table, had tea and the like, and then went up on deck. I watched them unload horses and cows from the ship. They bind the cows' front legs and lower them down onto boats. For the horses, they tie a rope around their waist and lower them down. I'd heard that the water is choppier here than elsewhere, but today it is perfectly still. Despite that, the boats are bouncing around and shuddering. Two of the women in my cabin departed, but another two arrived. They're both Russian. The Arab mem is still there though.

The ship only stayed here an hour before leaving. I spent the day either writing in my diary on the deck or coming and going from my cabin. In the evening I washed up. The dinner bell rang, and everyone went to eat. I didn't like the food and had to force myself to eat it. Then I went to the cabin to rest. Sarkar did the same in his own cabin.

[*December 11*:] I woke up as usual and went to have breakfast with Nawab sahib. They only serve toast and tea. There is no butter, jam, or fruit. You don't get enough to fill you up. The ship is dirty. I went to the deck to have a stroll. The Arab mem mostly stays in the cabin. I don't feel happy on this ship. The day passed. I had dinner and went to my cabin. So did Sarkar-e ala. We both slept. I should tell my readers that I refer to my husband, Nawab Sarbuland Jung Bahadur, as Sarkar-e ala.[27] The night passed, and another day ended.

At Port Said

December 12, 1909: In the morning, the ship was already at Port Said. We reached [the port] at 5:00 a.m. I woke up at six excited to finally escape this ship. I quickly washed up and got ready. Sarkar came, and we prepared our things and had our breakfast. The Arab mem boarded the small boat with me. Al-hamdulillah, God had at last delivered us from that disgusting ship! My attendant [Amina bi] was with us. I've forgotten the name of the ship, but it is recorded in Sarkar's diary. We both went to the station [port] in Port Said and had our luggage unloaded there. Once everything was arranged, Sarkar instructed a coolie to take Amina bi and our things to the train while we visited Port Said. It was a quick tour because the train was due to depart at 12:00 p.m. and we had to get back soon. After a quick tour we went to the Ashtaran Hotel and had a delicious meal.

After this rushed breakfast, we hurried to the station. Sarkar collected the tickets while I loaded Amina bi into the train with our luggage and then went to my own compartment. The train began to move just a few moments after Sarkar returned. He said that we would need to change trains during the journey and that we should go to Suez. We would need to transfer at Ismailia. We arrived at Ismailia Station two or three stops later and left the train. We got down, and Sarkar went to unload the luggage. I went ahead without him and retrieved Amina bi.[28] The second train was standing across the platform. First, I got Amina bi and the luggage into a carriage, and then Sarkar and I went to sit in our compartment. There is so much dust and sand in the air here that all the benches are covered in it.

At Suez Station

We reached Suez Station at 4:30. Al-hamdulillah. We both disembarked. This is a small town. There are hotels and such here, but on inquiry we were told that the port was far away. Sarkar said that we should stay near the port because he needed to make inquiries [for our onward travel]. We hired an excellent phaeton with two horses. Our luggage was loaded into a second carriage. We had Amina bi sit inside this second carriage and dispatched it to the port. Then we left in our phaeton, taking in the scenery along the way and viewing the Suez Canal. We reached the port in about twenty minutes and went to a nearby hotel. The hotel looked small but was in fact beautiful and luxurious, with "electric lights." It is clean and comfortable. We took two rooms: one for Amina bi and the luggage, the second for ourselves. In the evening we washed up and got dressed. We went to dinner when the bell sounded. The food was very good. Then we went back to the room and went to sleep. By the grace of God, the night passed uneventfully.

December 13, 1909: We woke up in the morning. After fajr we had breakfast. Sarkar went out to make arrangements. Meanwhile, I wrote and mailed letters to Hyderabad and Lucknow, to my pir, to my father, etc. Sarkar, having made our arrangements, came back around the time I finished my writing. We had breakfast.[29] The food was good. After eating we boarded the ship. [First,] we took a leisurely stroll along the sea toward the port. Sarkar had me sit in a room in the port building. There were many other rooms near it. Many other hajjis were standing outside. I showed my passport and documents. A doctor came and checked my pulse. There are many Turks. All the police are Turkish too. I sat down on a chair, observing everything. I kept my bag and umbrella close. Near my room, just five or so steps away, was a docked ship that looked like a train. It seems that the sea is deeper here, and the ground is higher. The ship was connected to the dock via a ramp with steps.[30] People went up and down. Many carriages, buggies, and camels were loaded. I heard that the day before yesterday the khedive of Egypt traveled to Jeddah and that they are now sending his luggage after him.[31] There were even flags strung around the port [in celebration of his departure]. The carriages are being loaded with great care and precision. I was told that the ship will leave at three. It took Sarkar about an hour to arrange everything. Then he came back.

On a Turkish Ship

We boarded the ship along with our luggage. The name of this vessel is *Khedive Oil Boat*.[32] It is a Turkish boat. Its crew includes both Muslims and Europeans. It is very clean. We are in first class. There are excellent purdah arrangements for women travelers in our class. This is the only ship I have seen with purdah arrangements for women travelers. It is a credit to Islam. While this ship cannot compete with P&O vessels, it is still very clean and comfortable. In any case, may God save us from the absolute filth of the Austrian ships.

We got settled in and arranged our things. Sarkar had one cabin. I stayed a short distance away in a cabin in the women's section. There are six beds in my cabin. The bathroom is very close and very good. I had Amina bi stay nearby. Sarkar's cabin has three beds. His cabin is just across from my own.

I took a sea bath. By then it was 2:30 p.m. The ship was still docked. I got dressed etc. Sarkar got ready too; then we went to the deck. There is a canvas canopy. Quite a few passengers here have simply rolled out their bedding right on the deck. They are all going for hajj. The ship departed but at 3:30 p.m. rather than at 3:00. A little while later the passengers began to do vuzu. They sound the *azan* at prayer times on this ship. Ah! What an Islamic blessing! As soon as we heard it, we both went back below deck to do vuzu. Sarkar did his asr prayers outside. I did mine in my cabin. Then I sat to write in my diary and continued with this until I heard the maghrib azan. After praying I went out for a little walk. As there is no one in first class but us, it is entirely empty, and I can walk around quite comfortably. Then dinner was served. The food was placed before me. I had heard that there were fully seven people in second class. May God have pity on the unfortunate people traveling in third class; they are numerous. We prayed isha and then returned to the table to eat. The food was excellent and carefully prepared. The captain of the ship ate with us. He is an elderly man with much experience and a Muslim. We had a long conversation on religious matters and took turns reciting verses from Qur'an. After dinner we went to bed.

A Stay near Mount Sinai

December 14, 1909: I woke and prayed fajr. Amina bi, the poor thing, is feeling very seasick today. Soon our ship anchored near Mount Sinai. They say this is the true Mount Sinai. The one I heard about in Nazareth is just using the same name. Moses once led the twelve tribes of his nation through these waters. Pharaoh drowned in this very spot. From this sea begins a road that leads straight to the Nile. We both read a fatiha for the Prophet Moses. The ship didn't stay here long. There are said to be many fruits here. They say that the place where Moses crossed is called Gharb in one direction and Zafaran in the other. Soon it was evening. I slept after isha.

December 15, 1909: Our ship stopped in the morning at a place called Lalluja. The khedive of Egypt's luggage, his camel carriages, etc., were being unloaded. There are motorcars too. They say that after coming back from Mecca he will travel to Medina from here. He has a lot of luggage with him.

I got ready and went to the table. Nawab sahib came too. We had breakfast and then went to the deck. Sarkar strolled in one direction while I went off in another. While walking, I saw a few Egyptian women sitting in third class. There is a door between our deck and theirs. I walked over to the door, opened it, and gestured to them so that I could have a conversation in Arabic. These poor but noble women came and sat just before the doorway and began to speak in Arabic. By now I can understand Arabic a little bit better than before. I examined their burqas and clothing attentively. They are also going to Mecca for the hajj. I couldn't learn much about them because I don't yet speak Arabic well enough. I took my leave from them and went to look out at the sea.

There are many fish here of a reddish, golden color. The people catch them and sell them, and then the passengers buy them and cook them. God, have mercy on the Muslim community, and may the Prophet help the Muslims! The place where these people stay is dirty and strewn with trash. It is the passengers themselves who have dirtied it. The ship's crew is lazy and disorganized. Our waiter does not clean our cabin or our shoes on time. The crew is surly too.

I spent the day watching all these little spectacles. Finally, the day ended. Dinner was served on time. Afterward, I went to my cabin to write up all that I had seen during the day in my diary. Then I slept.

[*December 16, 1909*]: We both woke up on our own time and did vuzu. We prayed fajr and then got dressed and went to the table to eat. When we went to the deck, we saw that the entire ship was being washed. Today is December 16, 1909. They say we will reach Rafi at five today, and then we must put on our ihrams.[33] The captain announced, "When we reach Rafi, I will sound my 'bugle.' This is the sign for you all to get ready." As usual, I went to the deck and stood by the railing, where I heard this announcement. Suddenly, though, my sari got caught in the railing, on a piece of metal that moved up and down. The fabric had gotten pushed inside a hole in the metal. I tried to extract it as soon as I realized, but it had already gotten entirely stuck. I tried to pull it toward me, but it wouldn't come out. Sarkar was far away, so I sat and waited for Amina bi to come so that I could ask her to fetch some scissors to cut the sari. I was still waiting when, by the will of God, the metal changed position again and the sari easily came out, though it was now completely soiled.

I went downstairs and changed. At two in the afternoon, the steward came and cleaned the cabin. The servants on this ship are a bit daft. This one spread out the rug at an angle. I asked him to please straighten it out, and his excellency replied, "No mind." It made me laugh. At 3:30 p.m. we bathed, put on clean clothes, and prepared to don our ihrams. The azan for asr began. Sarkar went to pray in the men's area. I prayed in the cabin. After asr the bugle sounded. We had reached Rafi. We all put on our ihrams. Then it was evening. We had dinner and then went to sleep.

Notes

1. Imam Husain (626–680) was the grandson of the Prophet and the third Shi'i imam. He was killed and beheaded in a massacre at Karbala. There are several claimed locations for the burial of his head, including in Damascus.
2. "Ma fi hammam ila hotel, lakin hammam fi al-balad."
3. To be used as a scrub or exfoliant.
4. "Marhaba, tafadhal, tafadhal."
5. "Asbar, asbar!"
6. "la, ujr."

7. “Ana Hindi. Al-Hind hamam al-bait, al-balad ma fi hammam lakin hina ana shuftuh hammam fi al-balad, al-bait ma fi. Ana ta‘jub.”

8. “Ya ukhti, hurmuhu hina.”

9. “Na‘m lakin ana baghit rumal.”

10. “Ma baruh, ma baruh, ajlas, ajlas, shuaya.”

11. “Ana ma asbar. Ana ‘ujlat. Ana al-jauz al-bab fi.”

12. “Enti al-Hindi tayyab?”

13. “Kathir tayyib.”

14. For an analysis of this and the preceding interaction, see Majchrowicz, *World in Words*, 170–72.

15. These refer to the four major schools of jurisprudence in Sunni Islam. By making available space for each school, visitors were able to attend lectures and discussions in accordance with their own legal affiliations.

16. John the Baptist.

17. A prophet, after whom the eleventh book of the Qur’an is named.

18. Khizr, or Khidr, is a mystical figure widely believed to possess great wisdom. He is believed to be mentioned in the Qur’an, though never explicitly.

19. “La, ba‘d zuhr raftani.”

20. “Bismillah, tafadhal, emshi.”

21. “Tafadhal, tafadhal.”

22. “Ta‘l ya sitti, nahnui nahnui hamam lazim.”

23. “Na‘m.”

24. An Ottoman coin.

25. The ship was Austrian.

26. “Ana araf arabi, la araf English, ma araf French.”

27. It is difficult to explain why this unexpected note—like a later one that introduces Amina bi—appears nearly halfway through her account, especially since the term *sarkar-e ala* has already appeared several times. Had it been an addition made during the editing process, it surely would have come earlier in the diary. One might speculate that, as she wrote, Begum Sarbuland began increasingly to think of her diary as a travel account that would be shared more widely than had perhaps been her first intention.

28. Amina bi was seated in another, presumably less luxurious part of the train.

29. This is not a second breakfast but a different type of meal. In the first instance, she writes having had *nashta*, and in the second, of having “breakfast.” The latter here clearly refers to an English-style breakfast, and the earlier nashta was a lighter meal perhaps including some type of Indian snacks.

30. Her vocabulary here is all borrowed from the language of train travel. She calls the dock a “platform” and the port a *jahaz ka isteshan*, or “ship station.”

31. Abbas Hilmi Pasha (or Abbas Hilmy II), the final ruler of Ottoman Egypt. He was deposed by the British in 1914. The khedivate was abolished, and Egypt was made formally into a British protectorate.

32. Sarbuland Jung gives *Rahmanieh*. It was built in 1865 and broken up soon after their trip. H. Khan, *Safarnama-e Madina*, 12.

33. Rafi marks the boundary after which pilgrims should enter a state of ihram, or purity, and put on the ihram garment prescribed for all pilgrims.

CHAPTER 6

MECCA

مکہ مکرمہ

The Holy Land of Arabia

December 17: Our ship reached Jeddah early in the morning. As is our habit, we woke up, did vuzu, prayed fajr, and got ready. When we left our cabin, we saw that the ship had anchored and that there was a heavy rain. We packed our things and were ready to get off the ship, but because of the rain, no boats came out to ferry passengers to shore. There was just one boat going back and forth. It was only three hours later, when the rain stopped, that the boats began to come and the passengers could disembark. The boats are very large. They carry fifty passengers at a time. We waited about until the crowd had reduced and left on the very last boat, which had fewer people. There were still twenty people in it, but we disembarked anyhow. These boats are called *baglas*. They have large sails and move with the wind. The boatmen shift the sail according to the wind's direction. The wind was still then, and the boat moved slowly. It was not raining any longer. We landed safely on shore after three hours. We thanked God upon our arrival in the city of Jeddah. Then we gathered our belongings and went to the home of Mahmud Asfi. He is an acquaintance of Nawab sahib's. This gentleman always allows travelers to stay with him. Our room is very nice. The floor is even covered with carpets. He also gave us good food to eat. We spent the night comfortably.

Traveling by Camel

December 18, 1909: We woke up and, al-hamdulillah, prayed fajr. We had already arranged our onward transport by camel. We had breakfast after the prayer. The camels and the shaghdafs were ready and waiting, but the sky was overcast. It was 8:30. Mahmud Asfi came and said, "My wife and the other [women] would like to meet you." I very happily accepted and went to the other side of the house and up the stairs to meet them. There were three ladies: his wife, his daughter-in-law, and his daughter. They speak Arabic. They gave me a warm welcome. I understood their language with great difficulty. I answered some of their questions and tried to convey my happiness and gratitude as well as I could. I then took my leave. We both, along with Amina bi, my servant, gathered up our belongings and left the house. My shaghdaf had already been placed on the camel. To enter it, I would have to climb a ladder that had been placed against the camel's neck. This was a frightening prospect. How can I climb up there? What if the ladder slips?

Nawab sahib has retained two Arab servants here. He told them that they would be employed for the entire hajj season. It was agreed that each of them would be paid two *ashrafis* in total. They were both thrilled at this and readily agreed. One is named Abd ur-Rahman. He speaks some Urdu. The other's name is Sayyidi.[1]

The ladder was leaned against the shaghdaf, and the two servants and a Bedouin stood nearby to support it. I took God's name and climbed up. Sarkar followed after me. We left Amina bi with Mahmud Asfi. "Please send our remaining luggage with Amina bi in the caravan leaving tomorrow," we said. Then, bismillah, our camel set off to join the caravan. Laden or not, camels move very slowly. They travel in a chain, with the rope of one camel tied to that of the next. This is a large caravan. There are nearly five hundred camels (fig. 6.1).

It began to rain as soon as the caravan departed. The wind was so fierce that it blew away the wax cloth that had been spread over the top of the shaghdaf. Soon the carpets that had been strung up around the shaghdaf to block out the sun were soaking wet, to the point that water began to drip inside. So much water came in that we were drenched from head to toe. The bedding was sopping wet too. The wind was so cold it felt like arrows piercing our bodies. It occurred to me that perhaps this was a blessing from God

Fig. 6.1 A caravan of camels mounted with shaghdafs, or litters, traveling to Mecca (1910). Begum Sarbuland did not enjoy traveling in this way. *Source: Library of Congress, Prints & Photographs Division, LC-M32-13770.*

and that it was for the best. The Bedouins were ecstatic. Their happiness was boundless. They said, "This caravan is a blessed one. It has not rained in this place for seven years!" We were so cold that our teeth were chattering, but because God's grace and kindness were with us, we didn't mind it at all. Along the way we didn't see the shade of a single tree nor any place one could make camp. The caravan pressed forward, with mountains passing us by on either side. If anyone said, "Brother, please stop the camel," the Bedouin would reply, "What? You want to be left behind by the caravan so that someone can see you alone and loot you?"

"All right, all right, let's keep moving then."

The rain continued for two full hours, but the caravan never stopped. Finally, thank God, al-hamdulillah, the rain stopped, and the sun came out, but the wind continued unabated. At least that had the effect of drying out my clothes. I could not remove my head covering because I was wearing an ihram, but Sarkar took off the upper portion of his ihram, which was soaked through, and gave it to Abd ur-Rahman to put on top of the luggage camel so that it would dry out. The clothes I was wearing eventually dried out themselves as I wore them. By God's grace, once the rain stopped and our clothes were dry, our moods improved. We weren't tired at all; in fact, we were hungry. We pulled the bedding from underneath us and had it placed on the second camel to dry.

Ahead of us we saw a few houses made of dried grasses. I was told that they sell coffee, that is, *qahwa*, here. One of our two servants ran to get two glasses for us. "These glasses cost one *hilil* each," he said, taking a little sip from it himself. I said, "What was that? Why did you defile it like that?"[2] He replied, "I am Muslim. You are Muslim. Same, same."[3] Sarkar drank the coffee, but I did not. The coffee cups are very, very small. These little coffee huts lined the path for several miles.

I had the shaghdaf stopped, and we got down to renew our vuzu while the caravan continued on in front of us. There are five hundred camels, after all. Then I mounted the camel again, now much more at ease. We have our food with us. We asked for the tiffin basket to be brought and opened it up. We both ate and gave the leftovers to the Bedouin. They were very happy. I had finally warmed up a bit. It was time for asr. We prayed asr in the shaghdaf itself. The camel driver and our servants were so pleased by the food we gave them that they hurried the camel back to the center of the caravan and tied it to the line, shooing everyone else out of the way. They kept saying to us, "Don't worry at all, be comfortable, we are here for you." They carry guns on their shoulders with cartridges strung around their waists. They take turns sitting on the second camel that is carrying our luggage.

We reached the end of the first stage at seven in the evening. This is a little village called Bahra.[4] There are many huts for travelers to use.[5] The shops are lit up with lamps. They sell food, sweets, and so on. There are charpoys and people drinking coffee. There are melons for sale in the shops. We made our way through this scene until we reached a plain. Here our camel was led away from the caravan through a plain to the door of one of the huts. There was a full moon. Its bright light illuminated everything, and we could still see the caravan. We got down from the camel. I was a bit tired. I went into the hut. Our bedding was still not completely dry. Despite this, I spread it out on the ground and, considering it to be dry enough, lay down.

Nawab sahib came in soon after. He was followed in by the Bedouin camel driver, who said, "O Sheikh! Baksheesh!"

"You can have it in Mecca."

Abd ur Rahman then said, "O camel driver! You will get your baksheesh in Mecca. Be content."

Then he left [the hut], but came back a few moments later and said, "Have some coffee." We drank it but did not like it because it had ginger in it. Then the camel driver said, "I had heard that you are a nawab and a begum. I was very happy because I assumed that I would be getting a lot of baksheesh."

I replied to him, "No, I am a *khanam*, not a begum."[6]

Afterward, I asked Abd ur-Rahman to explain to him, "We have nothing to do with the pomp and magnificence of nawabs and begums. We are nothing more than two slaves of the Prophet's door who have come to beseech God Almighty for mercy and forgiveness. We are neither nawab nor begum." He [the camel driver] was quiet after this. Then Abd ur-Rahman said to Nawab sahib, "Sheikh, you appear to have a lot of inherited wealth. This is how it seems to me." Nawab sahib just laughed, but I replied, "That is only because you hold us so dearly. In actual fact, we are nothing but base servants of the Prophet." Then Abd ur-Rahman went quiet too.

If the Bedouin find out that you are wealthy and have plenty of cash to your name, then they will loot you. But our Nawab sahib has organized things so well that we always have access to money, but we never carry with us a cent more than we need at the time. I was so afraid of the Bedouins that I did not let Nawab speak to them much. I was the one in charge of communicating with them, because Arabs take good care of women but like to fight with other men.

We slept for a while. After some time Abd ur-Rahman came and said, "Wake up, Sheikh."

"What is it?" we asked.

"We've agreed to a rate of two ashrafis, right?"

"Yes."

Then he went outside again. "Well," I said, "why is he thinking about that at such an hour?"

We woke up at 4:00 a.m., did vuzu, and prayed fajr. I had only just asked Nawab sahib to have the shaghdaf prepared, as usual, when Abd ur-Rahman began to call for Sarkar himself. Al-hamdulillah, after breakfast we got back into the shaghdaf. Sarkar stretched out comfortably, but I remained seated upright out of fear.

The Land of the Haram

December 19, 1909: In the morning the caravan resumed its journey to Mecca. At zuhr our Arab servants untied our camel from the caravan. We got down, did vuzu, prayed zuhr, and then remounted. They hustled the camel back to the caravan. Abd ur-Rahman said, "Sarkar, we are nearing the Haram now. You both should start to recite *labbaik* from here on."[7] We did as he advised and began to recite. The caravan entered the city of Mecca at asr. The people of Mecca may have had their fill of the sight of the House of God [the Kaaba], but those of us who thirst for a glimpse of it, we are full of passion, restless to lay eyes on it. I want nothing more than to go directly to the Haram, to see the House of God—the place to which I, by God's grace, have made this difficult, long journey—and press my eyes to it.

My God's Home

We were both ebullient as we went to our lodgings.[8] We immediately renewed our vuzu, put the servants in charge of our luggage, and rushed to the Haram with a tingling sense of anticipation. I cannot describe how my heart felt. I repeated, "Labbaik," over and over. My tears flowed freely. Soon I saw the entrance to the Haram before me.[9] How can I describe its celestial majesty? We rushed forward, increasing our pace with rising excitement until we reached it and kissed its doors. The Haram was there, immediately before us. At the Bab us-Salam our heads lowered in prostration. How enchanting is my God's house! Allahu akbar! Allahu akbar! Allahu akbar! How glorious is our God, who has honored us with the sight of this most pure, most blessed house. I fell to the ground in prostration spontaneously. My happiness and passion were indescribable. I prayed profusely for the entire Muslim community, for my parents, my friends and companions, for my children. I didn't want to raise my head from my prostration. I felt indescribable tranquility. My heart was entirely at peace. I was enveloped in a kind of happiness I had never felt before. I only raised my head again with the greatest difficulty. I opened my eyes and gazed upon the gloriousness of [God's] oneness. This was not just a house. It was God's light itself (fig. 6.2). *Al-azmat lillah*![10]

Tawaf at the Kaaba

The muallim said, "Come, now perform the tawaf." I stood up, still lost in my passion. I included the names of my parents in every prayer the muallim had me recite. *Ruhi fidak, ya Rasulullah*.[11] All of this has been made possible through you, through your intercession, by which we have had the good fortune to come here.

We began the tawaf. Ah! This is where the sky itself lowers down. This is where the holy land kissed the feet of Ruhi Fidak, the Master of the Two Worlds [the Prophet]. Some of the greatest prophets and saints once walked here. My happiness was ineffable. Again and again, I read: "O God! Accept us. We have come here at your command. You have honored us in every way. You are the Great Creator. All creation rests in the palm of your power. We are reliant on You at every moment. The heart that You placed inside me is here for You. Good or bad, we are Your servants. God, forgive us, through the intercession of Ruhi Fidak, the Prophet." My heart would under no circumstances agree to leave this place. The time for maghrib ended, and then it was time for isha. Soon we heard the azan—Allahu akbar!—and rose to pray.

Tears of Happiness

Near the Haram there is a place called the Hatim.[12] I prayed isha there and then raised my eyes to gaze upon that pure house [the Kaaba]. Then, in a state of hopefulness, I picked up its covering and wrapped it around me and then sat down, pressing myself against the Kaaba. I cried ceaselessly. Tears of happiness, excitement, and love rolled freely down my cheeks. Eventually I looked up again and saw that the moon had risen. The moon sacrifices itself for this pure house and bends down to kiss it. The view around me was even more enchanting in the moonlight.

Safa and Marwa

The muallim said, "Now you should run between Safa and Marwa." We went to this holy place, but there was a market there. This holy place is outside the Haram. There is a marker on one side [indicating Safa] and another at the further side [indicating Marwa]. One should read prayers as they walk briskly between the markers. This act is called "running Safa and Marwa." It seems to be about half a mile between the two signs. There are shops on either side of the course. It is run seven times. God on high had us perform this component [of the hajj] in a beautiful way. Then we went back to the Haram. We made several nafl prayers and then went to the Zamzam [well] to drink its water.[13] As I looked around in every direction, my heart swelled with elation.

We left the Haram. The muallim had us remove our ihram. We returned to our lodgings at ten in the evening, asked for food, ate, and slept.

Back to the Kaaba

December 20, 1909: We woke up at 4:00 a.m., still ebullient. We did vuzu and went to the Kaaba (fig. 6.2). We went through the Bab us-Salaam, and I went and sat in the Hatim. Nawab sahib joined the group prayers. We

Fig. 6.2 A view of the Kaaba and the courtyard of the Haram Mosque in Mecca at prayer time in 1889. *Source: Library of Congress, Prints & Photographs Division, LC-DIG-ppmsca-35470.*

prayed fajr according to the Shafii timing, which was happening just then, and then began to circle the Kaaba. We kissed the Black Stone from afar.[14] The crowds were so great that it was difficult to reach it, particularly for weaker people like us. Still, I firmly believe that one day God will make it possible for us to approach it and kiss it.

Soon daylight began to appear, and the Hanafi call to prayer began. Its sound awaked a sense of God's greatness in our hearts. After the prayer concluded, we resumed our tawaf and then ran between Safa and Marwa again before returning to our lodging. We had breakfast, did vuzu, and returned to the Haram. Here I learned that because it is now the eighth day of [the month of Zil-] Hajj we will leave for Mina today. The camel and shaghdaf are ready. Mahmud Ahmad, Sarbuland Jung's nephew, who had left Bombay on another ship, arrived in Mecca two days before us, traveling directly here via Jeddah. He is now staying with us.

Departure for Mina

We three left the house. The caravan was ready. We are also traveling with Halimullah's wetnurse;[15] Khuda Bakhsh; my servant Amina bi; Mahmud Ali, the steward; and our two servants.[16] They brought out everything we would need. We gave the servants one camel and one donkey. Khuda Bakhsh wanted to walk. Sarkar selected a very nice donkey for himself. Mahmud Ahmad rode in the shaghdaf. And thus, by God's grace, we all set off for Mina. We enjoyed the views as we went and arrived just at zuhr. Nawab sahib did vuzu and went straight to the large mosque. I joined him.

The khedive of Egypt is also on hajj. His being there has complicated matters, and there are extra arrangements being made to accommodate his presence. It was not easy for regular people to get permission from the Egyptian guards to enter the mosque, but, al-hamdulillah, we all got in. By chance, we found a space to pray near the khedive himself. We prayed with the congregation. After prayers I looked around and realized that I was the only woman there. The other women were somewhere else. After prayers Nawab sahib took me to my tent and said, "Maybe women don't come to this mosque. There was no woman there for prayers beside you."

"Very well," I replied, "I'll pray asr, maghrib, and isha in the tent. You go back to the mosque. My tent is very comfortable."

Nawab sahib returned to the mosque for asr. There is a wonderful arrangement here—namely, that at every prayer time the khedive's men fire their guns. I prayed asr in the tent. Then I went outside, only to see huge crowds and tents spread out in every direction for miles, as though I were in a large city. Every imaginable item from around the world is available for sale here; excellent foodstuffs are also available. All these items are imbued with God's glory. I took it all in for a while, feeling happy, and then went to my tent to busy myself in the remembrance of God. Readers [of this diary], please pray that God accepts my efforts, that they bear fruit, and that God have mercy on me.

Today is the ninth night of the month [of Zi al-Hijja]. In the morning we go to Arafat. I have heard that after completing the five daily prayers in Mina, after fajr, then you should set out for Arafat, but the crowds here are so great we cannot all go at once. For this reason, people begin to depart at midnight. Nawab sahib said, "You should also take Mahmud Ahmad and depart for Mount Arafat with our things at 2:00 a.m. Don't wait for the morning prayer. The crowds are too great, and you may be delayed.

You should reach there in time to hear the sermon." I agreed but replied, "I prefer to leave at 3:30 because the Shafii prayer begins at 4:00. My camel will surely still be within the bounds of Mina for the first thirty minutes of the journey. I can pray fajr on the camel itself. That way, I will have made my fifth prayer at Mina." We both agreed on this plan. Nawab ṣahib said, "I will go to the mosque at 3:00 and go straight to Arafat from there. Don't worry, an Arab servant will be with me. I will surely reach you in time, inshallah."

Departure for Arafat

On this evening I stayed awake until three at night in the most blissful state in the remembrance of God most glorious. The caravans began to leave. We decamped at precisely 3:30 a.m. Mahmud Ahmad and I shared the shaghdaf. The moon was setting, and we were waiting to pray fajr at four, fearful that we would exit Mina before then. Then, what do you know, it was four, and we both prayed while imagining the Kaaba before us.[17] Thankfully, I completed all five prayers at Mina itself. This was granted to me by God through the intercession of the Prophet. Now that I have seen the Kaaba, I have access to the simple expedient that any time I am traveling and cannot face it in prayer, I can imagine it before me instead. I see it in front of me the moment I close my eyes. God, grant this vision to every Muslim and every lover of the Prophet and his family! Amen.

After the fajr prayer, what did I see but clouds beginning to form in the sky. There was lightning too. This made me fearful, for I had not forgotten our last drenching. I prayed to God, asking him not to let it rain. I've enjoyed that enough, and it is the cold season. After praying I opened my umbrella inside the shaghdaf, just in case. Mahmud Ahmad laughed merrily at this and said, "Why is Mammani Jan [my dear aunt] opening an umbrella inside a shaghdaf?! It has a roof, after all." I replied, "Let me be, sir. We've already enjoyed one good soaking in this covered shaghdaf as it is. As they say, 'One burned by hot milk even blows on buttermilk.'"

We were still worrying about the rain when we heard the boom of guns being fired. We heard that this must be people firing in celebration, but Abd ur-Rahman realized that these were Bedouins who were molesting the caravan under the cover of darkness. He fired his own gun to scare them off. I was frightened. Why had the Bedouins come? Have they come to loot? We heard someone cry out, "A Bedouin has made off with one of the travelers' bundles!" This commotion continued until we finally reached the limits of Arafat safe and sound. Our tent was pitched near where the caravan from our nizam's city makes its camp, and where the yellow flag [of Hyderabad] was flying. When our tent was ready, I sat to rest until Nawab sahib arrived, safe and sound. Al-hamdulillah. He immediately took a bath. I wanted one too but found no opportunity to do so. At Arafat you are required to eat your meal before the sun reaches its zenith. We ate, as did the servants and Mahmud Ahmad. Afterward everyone did their vuzu and we busied ourselves with our *tasbihs* and *tahlils*.

The Day of Arafa

December 22, 1909 / 9 Zi al-Hijja: Today is the Day of Arafa. Nawab sahib, Mahmud Ahmad, and the servants went to the mosque for the sermon. Just one Arab servant stayed back with me. I learned that women do not go to the mosque here. They only go up to a certain point on Arafat and pray etc. By God's order, I too went to that blessed place with my muallim. Allahu akbar, this is the place where Adam and Eve were reunited. After a period of separation, God brought them together here. This mountain has been known as Arafat ever since; *arafa* means "to meet." This is a very powerful place for the acceptance of one's prayers. Prayers made here are immediately answered. I performed two rakats here and made many prayers. Then I spent a happy hour repenting for my sins. I finally rose to depart from this blessed place with unwilling feet; my heart did not want to leave. God! By the intercession of the Prophet and his family, forgive me, Nawab sahib, my parents, my children, all my near and dear ones, and every believer, male and female, and lead us down the straight path. Amen, amen.

As I returned, I saw what seemed like a vast city of people engrossed with all their hearts in the worship and praise of the Pure Protector, though some less fortunate souls were busy cooking, eating, arguing, and fighting (fig. 6.3).

The ideal course of action is to eat before the sun begins its descent and then to eat nothing more until you reach Muzdalifa. In any case, I returned to my tent. The muallim said, "We should leave here before maghrib. There is a place called Muzdalifa. We will go there and pray maghrib and isha together. We will stay the night there."

"All right," I replied.

At three in the morning, Sarkar sent me word that I should continue with the caravan according to my own preference and that he would go directly to the mosque at Muzdalifa on his donkey. "Once you've arrived and

Fig. 6.3 The pilgrim encampment at Arafat in 1916. *Source: Library of Congress, Prints & Photographs Division, LOT 3704, no. 78.*

had the tent set up, send the Arab servant to the mosque to find me. I will come back with him and rejoin you. Otherwise, it will be hard to find out where you have pitched the tent." I replied to say that this was a good plan.

I prayed zuhr and asr together and then thought to myself that I should go out and see what God's people are doing now. I went to the opening of the tent and looked out. What I saw made me realize: What do your actions matter to the Wantless God, tens of millions of whose servants are bowing down before him in prostration? You are too base to perform such worship. What will be your fate on the Day of Judgment? You are weak and no match for such good people. As I thought these things, tears began to flow from my eyes. I cried and cried until I was beside myself. I prayed. O Wantless God! By the intercession of the Prophet, by the sanctity of Mount Arafat, and that of your true followers, forgive this sinful one.

Departure for Muzdalifa

As the day slowly shrunk, and as more time passed, I thought to myself that this day was so blessed, and yet it was passing by so quickly. Ah! All that comes into the world must also leave it. This day ended too.[18]

The caravan was assembled before maghrib. I got ready. The tent was struck. The shaghdaf came, and I climbed in. Mahmud Ahmad rode with me. "God! I'm so sick of this shaghdaf. Please grant me some easier mode of transportation." After making this prayer I set off with the caravan just before maghrib. A few travelers' camels collapsed and their shaghdafs fell off, but they were all right. No one sustained any injuries.

The caravan arrived at Muzdalifa at isha. The moon had already risen. Everyone got down and removed their shaghdafs from the camels. Ah! Ah! Ah! Beauty was bursting everywhere. Nawab sahib had gone directly from Arafat to the mosque alongside the khedive and others.

I noted that no one was putting up their tents here. It was explained to me that people sleep in their shaghdafs here so that the tents can be sent ahead to Mina and assembled there before our arrival the following morning. Anyhow, after settling in I asked the Arab servant Abd ur-Rahman to go to the mosque and bring Nawab sahib back with him. "Take some food with you just in case he needs to stay late. If so, he can eat there." Then Mahmud Ahmad, the servants, and I all ate. I did vuzu and then spread my prayer rug behind the shaghdaf and made my prayers there. I presented my enfeebled state before my most merciful Provider. I was so very happy at this moment. Thankfully, my heart never forgot my parents at any significant point in this journey. I prayed for them first in every instance and only then for others.

When Nawab sahib came, he said to the muallim, "Good sir, where is my tent?" He then learned that "by custom, the tent has been sent ahead." It was very cold on the plain, so Nawab sahib said, "If there is any leftover tent, please pitch it because it is very cold. It is all right if not." The muallim went to investigate and found a tent that had been overlooked and left behind. This too is a sign of God's greatness that this wish of ours was granted. The tent was pitched. We went inside and busied ourselves in repentance, seeking forgiveness, reciting [the pious phrases] *la ilaha ila Allah* and *subhan Allah*. We fell asleep around midnight.

Al-hamdulillah. On the morning of December 23, we got up, prayed fajr, got dressed, and went outside. The caravan was ready. People were busy selecting pebbles. We were told that pebbles should be collected at Muzdalifa and taken to Mina to throw at Satan.[19] Nawab sahib and I both selected our pebbles, and then we set off. We traveled alongside the khedive's bodyguard and army. They formed a grand procession. Abd ur-Rahman led my camel away from the caravan and toward the procession so that I could watch it as we went. Mahmud Ahmad was on the camel with me. Nawab sahib is on the donkey. Guns are being fired because today is Eid ul-Adha. All of God's servants are wearing ihrams, even the soldiers. They look so very fine in their ihrams.

Entering Mina

We arrived in Mina at eight in the morning in high spirits, reciting labbaik and praising God all the way. I entered the tent. I was told that after the sacrifice, we should throw seven pebbles at the big Satan and then remove our ihrams. We will stay here until the twelfth [of the Islamic month of Zi al-Hijja], and every day we will have to stone the three Satans. On the twelfth we will throw our last pebbles and then leave for Mecca to do tawaf and run Safa and Marwa. I also learned that "on this day, if, after the sacrifice, one goes back to Mecca, circumambulates the Kaaba, and runs Safa and Marwa, they will earn great merit. However, they must return to Mina in time for the maghrib prayer. Staying in Mina for three days is a mandatory component of the hajj."

The Pull of the Kaaba

I prayed to my God, "O Lord, give me now the courage to do exactly this." After resolving to do it, I told Nawab sahib, "If you allow it, after praying and stoning Satan, I will go straight to the Kaaba, do the tawaf, and run between Safa and Marwa, and then return to Mina before maghrib."

He responded, "The Bedouins have taken all the camels away, and they will not return until the twelfth. How will you go? You could take my donkey, I suppose."

When I heard this, I became very happy. "That is wonderful," I replied. "If you give the command, I will go on your donkey. Abd ur-Rahman will accompany me, and if he wants, Mahmud Ahmad can too."

He gave his permission, and, at noon, after finishing all my tasks, I set out for the Haram. Abd ur-Rahman and Mahmud Ahmad walked alongside me. The donkey moved swiftly, and we reached the Haram an hour later. We immediately prayed zuhr, circled that pure house [the Kaaba], and ran between the Safa and Marwa. When I had finished with all this, I saw that the khedive of Egypt and his procession had entered the Haram Sharif too. I stood to the side of the haram and thought to myself: "Oh merciful, all-forgiving One! I rushed here excitedly because I believed that I would now be able to kiss the Black Stone while everyone is at Mina. I thought I would have it all to myself. Such is your glory that hundreds of your servants have come here to do the same thing. Today is Eid. Let my lips touch the Black Stone; after that do with me as you will.

> If you grant my wish, how wonderful; If you do not, I have no complaint
> My head is bowed in submission, I accept my lover's every whim[20]

This weak soul then thought: God! With the khedive of Egypt here, kissing the stone will be even more difficult. But then the seas of his mercy began to surge. A *siddi* of the African race wearing a turban suddenly came up to me and asked me the most wonderful question: "O my lady, do you want a kiss?"[21] I replied instantly, "Yes, sir, I do."[22] He immediately took my hand and pulled me into that massive crowd, protecting me as we went. It all became so easy. It was as though I were being carried along. Then, suddenly, he took my head in his hands and lowered it down to the stone. Then he stood over me, shielding me with his arms so that I would not be crushed. I kissed the stone again and again. Then he delivered me out from the crowd again. "Walk just behind me," he said. He spoke Arabic, but I understood him. He led me near to the *khatib* [area] and said, "Raise your hands [in prayer]." He made a lengthy prayer for my parents and me before handing me a piece of cloth from the covering of the Kaaba and a tasbih made of agate. I offered him two rupees, but he would not accept them. Instead, he rubbed his hands on my back and said, "Warm the soul. . . .[23] Now go."

I left the Haram Sharif, now very happy, and asked my servant to call for Mahmud Ahmad so that we could return to Mina on time. "The khedive is also returning now, so we won't be alone on the path." Abd ur-Rahman humbly replied, "If you would be kind enough to stop by my home for a short while, I will prepare a meal for you."

I replied, "Brother, we will be late."

"No, you will arrive in time."

Since he insisted, I accepted the invitation and went with him. His wife was already waiting for me. She excitedly welcomed me to her home and invited me to join her upstairs. There were a few other ladies as well. None of them could understand me. Abd ur-Rahman translated everything for them, and I also spoke [in Arabic] a bit. They served me excellent tea and gave me some fruit. Then I left. Mahmud Ahmad was just arriving. "I'm done now too," he said. "Let's go quickly now."

I got back on the donkey and left. The khedive was also just returning. It felt like we were returning together. There were huge crowds of people, but the path the khedive of Egypt took was a little less crowded, so I led my donkey in the same direction. The khedive and the sharif [of Mecca] rode ahead on horses while I followed along. I got a good look at the khedive. He is not old but actually quite a young man. He was moving slowly with his entire staff. I took it all in as we went.

Mahmud Ahmad was on one side of me, Abd ur-Rahman on the other. The latter started telling me his story. "I was with the raja of Nanpara when he came to Mecca. He stayed for eight months." Then he mentioned my father in Lucknow and said that he had supported him as a witness.[24] I stayed quiet as I listened. I didn't reveal the secret that I also had a connection to this story. I did not tell him because I had heard that if the Bedouin find out that you are rich or a nawab, then they will rob you. Even the Begum of Bhopal was nearly looted herself when she came for hajj.[25] The khedive is clever; he travels sometimes by motorcar, sometimes with his army. You can spend freely for godly causes all you like, but if you are discovered to be a nawab or a begum, you'll be looted. So I stayed quiet and listened to his story all the way to Mina. We arrived at the tent just in time for the prayer. They told me that Nawab sahib had gone to the mosque. I prayed maghrib in the tent and thanked God profusely.

By then I was exhausted. I lay down and had [a servant] give me a vigorous oil massage on my legs and back because I was so sore. Then I prayed isha. Nawab sahib returned with Mahmud Ahmad. We ate. Mahmud Ahmad made very good tea. Then we all went to bed.

Showering Stones on Satan

December 24, 1909: We woke up in the morning as usual and prayed fajr. We had breakfast and prayed zuhr early. Then we went to stone the Satans. *Oho!* There were hundreds of thousands of people here, all of them stoning Satan. Abd ur-Rahman and Sidi, our two servants, slowly pushed our way through the crowds until we were close to the Satan. I took careful aim and threw seven stones at each of the Satans before returning to the tent. Al-hamdulillah, I spent the rest of the day in remembrance of him. Nawab sahib said, "Today it is my turn to go to Mecca and perform the tawaf and run between Safa and Marwa." He left and came back right at maghrib. Then it was evening, and after eating and praying isha, we went to sleep.

In Mecca

December 25: Today the caravan will leave for Mecca after the stoning and the zuhr prayer, because the stones are thrown at the Satans after zuhr. We all had breakfast. Nawab said, "I am going to the mosque. Please do not wait for me. After you have prayed zuhr, throw your stones right away and then get into your shaghdaf with Mahmud Ahmad and set off. I'll come on the donkey."

"Very well," I replied.

I threw the stones after zuhr, and we set off. The caravan was very large and moved slowly. We reached Mecca only after asr. It was a challenge to go all the way home directly, so I got down [from the shaghdaf] some distance away and walked the rest. Nawab sahib arrived too. I took a moment to collect myself. Then I did vuzu, ate something, and left for the Haram Sharif. I did tawaf, ran the hills, and prayed maghrib and isha. Then I went home. Everyone had dinner. Mahmud Ahmad made tea and served it himself. We then went to sleep.

December 26, 1909: After fajr I had breakfast. Nawab sahib went out. I was also planning to go and see the city of Mecca. I took the muallim's wife with me and headed to the city. I saw the shops and such and bought a few things. We returned home for lunch. After eating we all did vuzu and went to the Haram Sharif for zuhr prayers. Then I returned home, where I prayed asr. I went back to the Haram Sharif again for maghrib and isha. After a tawaf I returned home, ate, and slept. We both went to the Haram Sharif at 1:00 a.m., where we joined God's servants in the tawaf. I kissed the Black Stone from afar, and then, after praying the morning prayer, I returned home to sleep.

December 27, 1909: I woke up at eight in the morning. I washed my face and did vuzu. After breakfast Nawab sahib said, "Let's perform an umrah today. To perform the umrah, we will need to go to a mosque that's far from here.[26] Once we are there, we will put on an ihram. We will then return, do a tawaf, run Safa and Marwa, and remove the ihram again." I said, "All right, let's go, but I will ride on a donkey. I'm afraid of the camels."

Nawab sahib called for two donkeys. We had food, got ready, and went outside. It was nine in the morning, and the donkeys had arrived. When I examined them, though, I realized that their reins were not like those on our previous donkey. There was just a rope tied around their neck. Riding in this way frightened me. I was worried I might fall. I refused. "Bring me a donkey with proper reins, or I will walk."

Our muallim, Siraj, said, "You should take a carriage." I was very happy to hear the word "carriage."

I asked, "Are there really carriages here? If so, I will certainly take one." I then began walking with Nawab sahib, and soon enough Siraj returned with a carriage.

This carriage is a strange sort of vehicle. It has two wheels and is shaped like a trunk. The rental costs six rupees. We got into it and set off. We were joined inside the carriage by one of the muallim's representatives, who provided guidance [for the pilgrimage] and who had come in his place. Our servant Khuda Bakhsh walked. I told the driver, "Brother, have Khuda Bakhsh sit in your place occasionally, and you walk beside us while holding the reins. Then you can switch places once again. In this way you can both take turns walking." I proposed this because the carriage looked quite rickety to me and because I had noticed that the driver would sometimes get out of the carriage while holding the horse's reins. He agreed to my proposal.

The cart driver drove the cart in a way that was beyond my comprehension. Although there was a clean, straight, smooth road already available, he led the cart up onto uneven, hilly, and rocky roads. He never stopped, whether there were potholes, water, or dust storms. There was no "coach box," so sometimes he would sit directly on the horse, or Khuda Bakhsh would. They took turns, but the horse itself was very small. The poor thing. First, it was attached to a carriage. Second, there was someone riding him. Third, it was being beaten constantly. Fourth, it was being made to climb up and down the roads, jostled about the entire time. I was vexed to see how this carriage was being driven. If it toppled over, there was just a small door to escape from. It would be difficult to get out. I was petrified all the way to our destination, which, al-hamdulillah, we reached safely.

When we entered the mosque, we saw that many people had come there to begin an umrah. In front of the mosque there was a large stepwell with a staircase. Everyone was doing vuzu there. We did ours and prayed two rakats. We put on our ihram. Everyone left one by one. We also departed, riding in the same carriage as before. This vehicle is called an *arabiya*. Now the driver drove the horse even harder than before and ignored us when we protested. Even worse, he told Khuda Bakhsh to sit on the horse and then climbed up behind Khuda Bakhsh himself. Now the horse had two riders, and it tried to flee. We had hardly covered a quarter of the total distance when the cart fell over and broke a wheel. Al-hamdulillah, no one was hurt, not even the horse. Now the question was how we were going to continue, as the distance was still great. The sun was very intense. The driver said, "I will go right now and get another horse cart; just wait here." Then he hopped on a pony and fled the scene. We waited for a while, and then I said, "Well, I'm going to just walk until he returns. At least I can cover some of the distance." I set off. Nawab sahib, the muallim, and Khuda Bakhsh came with me. There was a coffee shop along the way. Sarkar said, "The sun is very strong, and we are wearing ihrams. I cannot continue walking in this way with my head uncovered. We only have one umbrella, and only one of us can use it at a time. I will continue when the cart comes or if we can find a donkey."

"Very well, you ride a donkey. I'm going to walk because the donkeys here don't have reins."

I began to walk. Khuda Bakhsh came with me. I opened the umbrella because the sun was very strong. Slowly but surely, that blessed place began to come into view and my heart grew increasingly joyous. Everything here gives the heart a sense of tranquility and contentment. You feel peace and pleasure. It is a mark of God's glory that the fatigue and vexation I feel disappears the moment I place my hands on the covering of that pure house [the Kaaba]. God! Please allow my children, my parents, and all Muslims everywhere to come here quickly and visit all these places. Amen, amen. And please forgive me, in the name of the Holy Places.

I made these prayers as I walked. There were many donkeys on the path, but none of them were available for rent. After I had gone a long way, I heard an Arab cry out, "Transport, transport!"[27] I stopped him and said [in Arabic], "Me the husband in coffee-khana. It want transport. You go from here. After it, pocket [come]. It coffee-khana by umrah mosque."[28] I tried my best to explain my intention to him in Arabic using the words I had picked up from other Arabs during the journey. Once he understood me, he said, "You go little, you look, you me go in transport."[29] What I had said was, "My husband is at the coffee shop that is by the umrah mosque. He needs transportation. You please go and collect him." And he had replied, "All right, you come with me on the donkey and inform him yourself."

I then said, "No, I wait here and you your pocket," meaning that I would stay in this place until he came back.[30] He left immediately. I saw a coffee shop, so I took a seat and paid two *hilhils* for a cup of coffee. Then what do I see but Nawab sahib coming down the road on foot. The transport had not found him. We decided to catch our breath before continuing, when suddenly the original cart driver appeared with a new cart. I said, "I'm done with this cart. I'm already close to the Haram Sharif anyhow." Sarkar insisted, though. "No, no, please sit. He will be glad if you do." I got inside as before, and the driver began to push the horse as relentlessly as earlier. In the end, we made it to our destination, al-hamdulillah, and emerged from the cart unscathed. We entered the Haram and prayed two rakats on entering the mosque before our tawaf.

After the tawaf Sarkar said, "I'm getting hungry. I will go home to eat before running Safa and Marwa. Then I'll remove my ihram. You come too. You haven't had anything since breakfast and its already two thirty."

"You go ahead and eat," I replied. "I will run Safa and Marwa first and then remove my ihram and come back so that I can be comfortable and have a relaxed meal."

Then Nawab sahib went to our accommodations while I went to run the hills and complete this requirement. Then I headed home. On the way I bought a melon and carried it with me. When I arrived, I learned that Nawab sahib had just left for Safa and Marwa.

Then a representative of the muallim, a youngish man, came and opened my ihram. He cut my hair, made a prayer, and left again.[31] Much contented now, I washed up and ate some melon. I was happy. Then I had food and napped until asr. I woke up and did vuzu. I was told that "the asr prayer has already finished at the Haram," so I prayed at home. A while later I went to the Haram. The azan began when I entered. Allahu akbar! Aha! What a beautiful, enchanting, soothing sound! The prayer began right then. I stayed there until isha. I did the tawaf, and then we both went home. We ate and went to sleep but awoke and returned to the Haram at two in the morning. Allah! Allah! Even at this hour, there were still hundreds of people doing the tawaf. The tawaf continues day and night. It is such a wondrous thing. Do these people not sleep at all? The Haram is always full, a testament to God's glory. We did tawaf and then went to recite God's name in the Hatim. I placed the covering of the Kaaba over my head and knelt in prostration before the Provider, revealing to him my wretched state.[32] The peace, tranquility, and happiness I felt here is beyond description. I remained this way until it was the Shafii time for prayer. The cry of "Allahu akbar" reached my ears, and we both then prayed fajr according to the Shafii timing. Afterward, I sat and clung to the covering [of the Kaaba] until the sun came up. Then we returned home and rested for a while.

Visiting the Prophet's Place of Worship

December 28, 1909: We had breakfast at seven. Nawab sahib said, "Let's go pray a fatiha at the Jannat ul-Mala." "I'm ready, let's go," I replied. Once we were both ready, we left and strolled at a leisurely pace. First we had to climb the *jabl*, or hill. This place is very high, like a mountain. We climbed it. There is a mosque at the top which is said to be where the Sultan of the Two Worlds, the Beloved Prophet, used to worship. In addition to the mosque, there is also a place on the jabl where the Lord of the Two Worlds is said to have received messages directly from God.[33] We performed two rakats and said prayers in both of these places. We prayed that God would allow all Muslims to make a pilgrimage to this place once in their lives. I collected a few stones here as a keepsake. Then we descended the mountain via another route from the one we had ascended. There are a few houses and settlements here but not many. The place is so high that you can see the entire Kaaba and the city below. We easily descended and went straight to the Haram for zuhr. The azan sounded just as we arrived. We prayed and then returned home.

After lunch Nawab sahib said, "I need to go and meet Maulana Abd ul-Haqq to obtain an *ijazat* for *Dala'il al-Khairat*. I will go to him directly from here. Let's postpone the visit to the Jannat ul-Mala until tomorrow, inshallah.[34] We don't have much time left [here]." I agreed.

I prayed asr at home and chatted with the muallim's wives in Arabic. I prayed maghrib and isha in the Haram, came home, had food, and went to sleep. As usual, I spent half the night at home and half in the Haram Sharif.

Pilgrimage to the Jannat ul-Mala[35]

December 29, 1909: I had breakfast, did vuzu, and got ready to go. Sarkar said, "Let's perform the remaining visitations today." I agreed, and we went out, walking slowly on the holy ground of Mecca. It gave me a strange sensation to realize that this was the very ground that our most Beloved Prophet had once trod. We walked until we arrived at the Jannat ul-Mala, which contains the graves of many saints and great men of the faith. We prayed fatihas over them all.

The Prophet's Birthplace

Then we visited the birthplace of our Beloved Prophet. We passionately pressed our eyes to this special place, kissed it, and prayed two rakats. Then we went to the Mosque of the Jinn, where we have heard that the Master of the Universe taught the Qur'an to the djinns. We prayed two rakats here too. Then we went to the blessed place where the Prophet of God once lived and where the Lady of Heaven, Bibi Fatima uz-Zahra the Pure, was born. Ah, this is a place worth seeing! We both prayed two rakats. While we were in the house, a kind individual showed us another place. He said, "This is the very place where the Prophet would prostrate in prayer." Overcome with longing and devotion, we kissed this place and made our own repeated prostrations there, one after the other. Mashallah! How greatly has God blessed us to have been able to prostrate in the very place where the Lord of the Universe, the Master of the Two Worlds, once did! May the Prophet, through God, raise us up and raise up all Muslims. Then we went to a *beri* [jujube] tree that is said to have been here since that era. It still bears small fruits and is fruiting even now. I looked at it and all around it in astonishment. Before leaving, I collected a few jujube fruits.

We then went to the place where they say that Hazrat Ali ibn Abi Talib was born.[36] *(Although this is completely false. The Commander of the Faithful was born inside the Kaaba, as is clearly demonstrated in books using both Shia and Sunni sources.)*[37] We prayed a fatiha and two rakats here. After finishing our visits to all these places, we returned to the Haram because it was nearly zuhr. We arrived just as the azan began. We all prayed. After the prayer Nawab Sarbuland Jung Bahadur said, "I will go and meet Maulana Abd ul-Haqq directly instead of going home for lunch. I have some business there. You head home." I went back, ate, and rested for a while. Nawab sahib came home and ate. Then we both did vuzu and went to the Haram for asr. After the prayer we stayed in the Haram, gazing at the Kaaba and listening to the conversations of the people around us. There was such a wonderful atmosphere there. Then it was time for maghrib, and everyone stood. Al-hamdulillah, we prayed and did tawaf. Today I performed one tawaf each for my parents and then prayed isha.

God's Kindness toward His Servants

I forgot to write that when we had finished praying maghrib, it suddenly began to rain. Everyone rushed to drink the rainwater falling from the roof of the Kaaba. The place the water falls from is called the *mizab-e rahmat*.[38] My heart began to thirst for that water, but I thought to myself, "How will I manage to drink it when there is such a crowd?" I raised my hands and prayed, "O God! Please give me this water to drink. I cannot manage it myself." Just as I finished my prayer, a boy came running toward me with a cloth soaked in this water. "O mother! Do you want this water?"[39]

"Yes," I said, gushing with joy, "I want water."[40]

He wrung some water out of the cloth into my open palms, and I drank it. My heart lit up. I was overjoyed. Just see how quickly God has answered my prayers. Such is God's glory!

After the tawaf we went home, ate, and slept.

Milad Sharif in the Haram

December 30, 1909: In the morning we prayed, washed up, and ate. "Today is Friday," Nawab sahib said. "I've arranged for the Maulud Sharif to be read in the Haram today. There will be sweets and such and cooked food, like last time. Everyone who participates in the Milad Sharif will be fed. The cooks have all arrived, and everything is arranged perfectly."

The food and sweets were prepared by eleven. After praying a fatiha, the food was loaded onto trays and sent out to be distributed. Some was given to mendicants and the poor. We brought the sweets, which weighed about two *man*, with us to the Haram.[41] The Maulud Sharif was held after the Friday prayer. Al-hamdulillah, it was all done very well. Many people attended, and the sweets were all distributed. When we went home, we also fed the people who went there with us. All of us ate very well too. The food was delicious, peerless. The cooks and chefs were given one pound for their work. A few Arab ladies approached me. I gave them a meal and some sweets. One of these women gave me an Arab outfit she had sewn for Qudsiya.[42] I thanked her and gave her five rupees. We were still at home at maghrib. I took my leave from the ladies, and they left. I quickly got ready and went to the Haram to pray maghrib and isha. I did tawaf before going home. Everyone ate and then went to sleep.

Today is December 31, and, sadly, due to certain circumstances, I was unable to go to the Haram today.[43] Nawab sahib went himself while I spent the day getting ready and packing our things, for the time of our departure is near. Ah! How sorrowful it is that this beautiful era has come to an end so quickly. I spent the day reflecting on this. Nawab sahib bahadur came home in the evening. We all ate and then slept.

January 1, 1910

I spent the day at home again. Sarbuland Jung bahadur went out after eating. Some Arab women came to ask me for alms. I gave each of them something from what God has provided me. I thought that today would be a good day to go to the bazaar to buy some gifts, and so I went out, taking the muallim's wife with me. Christians sell goods at some of the shops here, [but] I went specifically to shops owned by Arabs and bought some Arab-style clothing for the children.[44] I bought some scarves for Nawab sahib and some gifts for my siblings. The bazaars here are covered. When I had finished, we went home. Nawab sahib said, "We will leave the day after tomorrow, on January third. Please prepare." This news unsettled my heart. How can our pilgrimage to this place have ended so soon? In any case, the day passed, and at night, we slept.

Separation from the Land of the Beloved

January 2, 1910: I spent the day meeting with various people. Nawab sahib went to the Haram and spent most of the day meeting with people as well. Then it was night. I spent the night sitting and gazing at the Haram. Ah!

What thoughts passed through my mind! It was such a sad thing that, due to various reasons, I could not go present myself at the Haram.

Right now it is seven in the morning on the third of January. We are all ready. The camels have come. We parted from Khuda Bakhsh and Mahmud Ahmad here because they will go to Medina. We have brought the servant Mahmud Ali with us. From among the women, we are bringing Amina bi and Halimullah's nurse with us. It is a pity that, for various reasons that have arisen, the servant Mahmud Ali and Halimullah's nurse are unable to visit Medina. Amina bi has already been there with me. At any rate, our things were loaded up. The servants took their seats, but I wanted to walk, at least for a while. This is the street of my beloved. The more I honor it, the better.

The caravan appears to be immense. Sadly, many people are returning directly to India without visiting Medina because they cannot afford to pay for the journey. Nawab sahib and I walked to the entrance of the Haram. He went inside while I prostrated myself outside the entrance of this sacred space before reluctantly parting from it. For a while some of the people we had met during our time here walked alongside us. Eventually, though, our shaghdaf emerged from the narrow streets safely. We both sat inside, and all of our friends returned home. And then, alas, the Land of the Beloved slowly but surely slipped from view. How could this happen, after all we have seen?

We had a pitcher of Zamzam water with us. Drinking from it steadied our tumultuous hearts. Nawab sahib's eyes glistened with tears. While we were sorrowful, we were also overjoyed that God had, in his great mercy, brought us to his door. This is no small honor. We hope that that Pure Being, our Provider, will grant us further mercy still and summon us to come here again, accompanied by our parents and children, with honor, prosperity, wealth, and health. We set off with this wish firmly lodged in our hearts. We continued until night began to fall. The caravan pressed forward without lighting any lamps. There was no moon, and the night was getting darker. Still, these Bedouins continued confidently on into the night, until there was nothing but darkness. The caravan stopped when we reached Bahra. We dismounted and went to sleep in the huts. We passed the night there just as before.

January 4: We woke up at five and had the baggage loaded back onto the camels. I don't feel well. I didn't digest yesterday's food properly, and I have developed a temperature. My entire body is aching. Still, I set off. Eventually the day broke. Suddenly, there was a sandstorm the likes of which I have never heard nor seen before. The sand dunes were completely blown away, only to form again elsewhere. The grains of sand flew like arrows. The sky and the hills were blotted out, and sand filled my hair and pushed its way into my eyes and throat. Despite the intensity of the storm, the caravan lumbered on. The Bedouin did not stop. Camels often die on the way. When one does, they just leave it where it falls. The storm continued for another five hours and only began to relent as we neared Jeddah. Al-hamdulillah, we finally reached our destination. It was 2:30, time for zuhr.

Sarkar got down [from the camel] the moment we reached Jeddah. He had Amina bi sit in his place to keep the shaghdaf balanced and then left to make arrangements for our sea travel, in the hopes that we would be able to depart today itself. My servant Abd ur-Rahman had my camel removed from the caravan. He then dragged it toward our house, fighting all the way, so that I would arrive there quickly. He spent a full two hours struggling with the camel before we finally reached Mahmud Asfi's home, as before. I am feeling very sick now and have a fever. I did not eat anything, on account of my upset stomach.

Instead of having a meal, after going into the house I asked for two pomegranates, had the seeds removed, and ate them. Though it was a struggle, I wrote two letters to Hyderabad and had them posted. Nawab sahib was still making arrangements. Amina cooked *qulia* for me. The poor thing went to the market to buy the groceries herself. I stayed in bed. Nawab sahib came and said: "By chance, there is a ship headed to Constantinople, the *Assouan* [Aswan]. It belongs to the Khedivial Company. Its departure was delayed by today's storm and will now leave tomorrow. If it weren't for this ship staying back a day, we would have had to wait a full week for another, maybe even longer. People arriving in the caravans from Mecca often must wait as long as two weeks in Jeddah Sharif. It turns out that the storm ultimately worked in our favor." Sarkar bought tickets on this vessel.[45] Then it was evening, but I couldn't eat anything. The fever did not go down. I passed the night fitfully.

Notes

1. Begum Sarbuland's attention to these personalized details distinguishes her account from those of her husband. Here, she gives the names of the two men, while he records them as "unknown Arab servants." H. Khan, *Pilgrimage to Mecca*, 52.

2. "Tum ne jhuta kyuñ kiya hai?" The term *jhuta* refers to food that has been made impure by being touched by another person, often of inferior status or caste. The servant suggests that, as fellow Muslims, his touch should not be defiling, but she does not agree.

3. "Ana Islam. Enti Islam. Sawa sawa."

4. A location about midway between Mecca and Jeddah. It was a traditional stopping point for camel caravans between the two cities.

5. Hamidullah records his disgust with and discomfort in these accommodations at length. H. Khan, *Pilgrimage to Mecca*, 54.

6. "La, ana khanam, ana ma begam." As noted later in her diary, the term *khanam* was current in the Ottoman Empire as a mark of the Turkish aristocracy. By invoking it here, she seems to have wanted to differentiate herself from the idea of opulent Indian rulers. Particularly, she may have thought that distancing herself from the word *begum* at that moment might have removed any assumed association with the fabulously wealthy female rulers of Bhopal State, who were known as begums. However, her effort here seems not to have been very successful. Several of Bhopal's begums had performed pilgrimages, including Begum Sultan Jehan most recently in 1903.

7. "I am present," a ritualistic phrase chanted by pilgrims arriving for the hajj.

8. They stayed in one of the "nizam's rabats." This was a collection of buildings purchased by the nizam of Hyderabad in 1860 to provide housing to pilgrims from the state. Though many of these buildings have since been demolished, this charitable establishment still houses several thousand pilgrims from the regions once ruled by the nizam today. H. Khan, *Pilgrimage to Mecca*, 55.

9. The Haram Sharif, also known as the Masjid ul-Haram or the Great Mosque of Mecca, is a mosque that surrounds the Kaaba. It encompasses several important sites and hosts many of the hajj's most central rituals.

10. "All greatness belongs to God."

11. "My soul is devoted to you, O Prophet of God."

12. The Hatim is a low wall that is set directly across from one side of the Kaaba. It is located not near but almost at the center of the Haram.

13. Zamzam is the name of a venerated well that is located inside the Haram.

14. A black stone set in the corner of the Kaaba, it is revered as a relic and said to have been preserved since the Creation. It is often kissed by the devout, but its small size means that few can get close enough to do so. Thus, they "kiss" it from afar.

15. Halimullah was the author's son who was born in October 1909. Strangely, Halimullah is nowhere mentioned earlier in the diary. The appearance of his nurse here in Mecca is similarly difficult to explain. It seems that a large contingent came directly from Hyderabad to join them on the hajj, including this nurse, who, based on her inclusion here, was probably a longtime employee of the family. She seems to have brought the infant Halimullah with her and returned him to Hyderabad afterward.

16. Nawab Sarbuland does not list his fellow travelers here but instead refers to them only as "our Indian servants." Amina bi is also not named in this section but only referenced as "a servant."

17. That is, they imagined it before them rather than facing it physically, due to the constricted movement inside the litter.

18. An Islamic day ends at sunset, or maghrib.

19. The Stoning of the Devil is one of the rituals the comprises the hajj pilgrimage. It involves throwing a total of forty-nine stones at three pillars that symbolize Satan and temptation. The stones are traditionally collected in Muzdalifa.

20. "Agar bakhshe zehe qismat, na bakhshe to shikayat kya / sar-e taslim kham hai, jo mizaj-e yar meñ a'e." This is one of the few instances in which Begum Sarbuland cites poetry in the diary. The verse is attributed to Nawab Ali Asghar (b. 1860).

21. "Ya sitti, anti baghit bosa?"

22. "Na'm ya Sayyadi."

23. "Ɛukhan ruh minaniha lama." The meaning is unclear.

24. It is unclear what legal case her father, Nawab Agha Mirza, might have been involved in.

25. This is likely a reference to the famous travel account of Nawab Sikandar Begum, which was originally written in Urdu but published in English. See Sikandar Begum, *Princess's Pilgrimage*; Sikandar Begum, "Tarikh."

26. They traveled to the Masjid ut-Tan'im, or Ayesha's Mosque. H. Khan, *Pilgrimage to Mecca*, 74.

27. "Markab, markab."

28. "Ana al-jauz fi al-qahwa khana. Hada baghit markab. Enta ruh min hina, ba'dain jaib hada. Hada qahwa khana qarib al-masjid 'umrah." The word "jaib" here is likely intended to be ja', "come."

29. "Enti ru suaya suaya ana enti shuf ana enta ruh min markab."

30. "La ana asbar hina enta jaibkum."

31. A ritual act and a mandatory component of the umrah and hajj.

32. The Kaaba is covered in a large, black curtain.

33. Hamidullah's travel account says that this was the Jabal Abu Qubays. H. Khan, *Pilgrimage to Mecca*, 72. He also records visiting several other locations that she does not mention.

34. *Dala'il al-Khairat* is a canonical collection of prayers composed in the fifteenth century. An *ijazat* is often a certificate confirming mastery of the text. Abd ul-Haqq was from Allahabad and a friend of Hamidullah's father. See also page 254.

35. This is presumably the Jannat al-Mu'alla, an important cemetery in Mecca that contains the graves of many important figures from Islamic history. It is sometimes called the Jannat al-Ma'la, but Begum Sarbuland's travelogue consistently gives it with a central *alif* rather than an *ain*.

36. Ali ibn Abi Talib (600–661), or Ali for short, was the Prophet's cousin, the fourth Sunni Caliph, and the first Shi'i imam. Begum Sarbuland here includes a long list of epithets for him that I have omitted.

37. This parenthetical seems to be an addition made by the book's editor or scribe. In this case, the insertion aims to pose an alternative to Begum Sarbuland's claim about the location of Ali's birthplace.

38. Literally, "the water-channel of mercy."

39. "Ya Ummi, enti turid hada ma'?"

40. "Na'm, ana urud ma'."

41. An immense amount of sweets, weighing approximately 160 pounds!

42. The author's second daughter.

43. She had presumably started her period.

44. Here, as elsewhere, Begum Sarbuland seems to believe, incorrectly, that all Arabs are Muslim and that non-Muslims are not Arab.

45. Hamidullah's account gives more details here: "With some difficulty I reached the office of the Khedivial Steamship Company. The agent had gone to start a steamer; but the sea being rough and stormy, he was not able to reach the ship or give orders for its starting. We had a return passage by the mail boat from Jeddah to Suez and our many passports fully entitled us to return to Egypt, but I was told by the agent's clerk that they had received strict orders from Government to allow only Egyptian subjects to embark for Suez. He offered to book us by the S. S. *Assouan* to Constantinople. We accepted the offer, not desiring to stay in Jeddah. But for the providential storm, the *Assouan* would have started today." H. Khan, *Pilgrimage to Mecca*, 76.

Plate 1 The cover of Begum Sarbuland's travelogue, *Dunya Aurat ki Nazar meñ*. The third extant print copy of the book bears a red rather than a green cover. *Source: Daniel Majchrowicz.*

Plate 2 A photograph of Begum Sarbuland Jung's father, Nawab Agha Mirza Beg Sarvar ul-Mulk. Probably late nineteenth century. *Source: Daniel Majchrowicz.*

Plate 3 Studio portrait of Begum Sarbuland's father, Agha Mirza Beg, with son and son-in-law, probably late 1890s. Seated, L to R: Agha Mirza Beg (Sarvar ul-Mulk), Nawab Hamidullah Khan Sarbuland Jung. Standing: Either Zulqadar Jung, the eldest son of Agha Mirza Beg, or his second oldest son, Jivan Yar Jung. *Source: Jawahar Khwaja.*

Plate 4 A portrait of Begum Sarbuland's mother, Sikandar Zamani, perhaps around 1920. *Source: Abid Ilahi.*

Plate 5 A portrait of Nawab Hamidullah Khan Sarbuland Jung, probably early 1900s. *Source: Jawahar Khwaja.*

Plate 6 A portrait of Nawab Hamidullah Khan Sarbuland Jung taken by the famous photographer Raja Deen Dayal, probably late 1890s or early 1900s. *Source: Abid Ilahi.*

Plate 7 A studio portrait of Begum Sarbuland and her family members in London in 1910. Standing, L to R: Her son Mahmudullah, her future son-in-law Abdul Majeed Khwaja, and her brother Dr. Sajjad Beg Usman Nawaz Jung. Seated, L to R: Begum Sarbuland Jung and her brother Mirza Abbas Beg. *Source: Abid Ilahi and Taj Khwaja.*

Plate 8 Begum Sarbuland with children and a nurse circa 1915. Standing, front row, L to R: Masudullah, Halimullah, Tehzib Nur, and Maqsudullah. Seated, L to R: Khurshid Begum, Begum Sarbuland Jung. Standing, back row: Nurse holding Usman Abida Sultan. *Source: Saberah Malik.*

Plate 9 Begum Sarbuland with her children, nephew, niece, wedding guests, and a bride, probably taken in Hyderabad circa 1920. Standing, L to R: Son Maqsudullah, Mirza Qulich Beg (future husband of Tehzib Nur and father of Shakira Sultana), unknown man and boy. Seated on chairs, L to R: Begum Sarbuland Jung and the bride. Seated on floor L to R: Begum Sarbuland's niece Akhtar and her youngest child, Usman Abida Sultan. *Source: Abid Ilahi.*

Plate 10 A signed portrait of Begum Sarbuland's oldest son, Muhammadullah (Mahomed Ullah). *Source: Sahar Mahjabeen Hasan and Shakira Sultana.*

Plate 11 Portrait of Begum Sarbuland Jang, probably 1920s. *Source: Najeeb Jung.*

Plate 12 Begum Sarbuland with grandchildren, possibly late 1920s. Standing L to R: Khalil Jung and Sami Jung, sons of Muhammadullah. Seated on chair: Begum Sarbuland Jung with newborn grandchild. Seated on floor: Razia Sultana, eldest daughter of Qudsia Sultana; Razia Beg, daughter of Akbar Beg, brother of Begum Sarbuland; and Fazal Ali Akbar, son of Qudsia Sultana. *Source: Saberah Malik.*

Plate 13 An undated portrait of Begum Sarbuland's eldest daughter, Khurshid Begum, possibly late 1920s. *Source: Sahar Mahjabeen Hasan and Shakira Sultana.*

Plate 14 A photograph of Begum Sarbuland's daughter Usman Abida Sultan with her nieces (the children of her elder sister, Khurshid Begum), probably in the late 1920s. From L to R: Alima Khwaja, Usman Abida Sultan, Taj Khwaja, Sikander Khwaja, and Amtul Jameel (Munoo) Khwaja. *Source: Abid Ilahi.*

Plate 15 A photo taken during Begum Sarbuland's son Maqsudullah's wedding in 1929. Standing, L to R: Khalil Jung next to his father Muhammadullah, Taj Khwaja, Abdul Majeed Khwaja, Mahmudullah, Akhtar Khwaja. Seated, L to R: Usman Abida Sultan, Begum Sarbuland Jung, bride Muhammad Zaman (princess of Kurwai State), groom Maqsudullah Jung, Khairunnisa (princess of Mangrol State, wife of Mahmudullah), Tehzib Nur. Seated on carpet, L to R: Sami Jung (son of Muhammadullah), Asif Jung (son of Mahmudullah). *Source: Saberah Malik.*

Plate 16 A family photograph of Khurshid Begum's (Begum Sarbuland's daughter's) family, possibly late 1920s or early 1930s. Standing, L to R: Akhtar Khwaja, Taj Khwaja, Habiba Khwaja, and Amtul Jameel (Munnoo) Khwaja. Seated, L to R: Abdul Majeed Khwaja, Sikandar Khwaja, Khurshid Begum holding Raveend (Rasheed Bilal Yusuf) Khwaja, and Alima Khwaja. Seated on floor: Ahmad Jamal Yusuf (Amannoo) Khwaja. *Source: Jawahar Khwaja.*

Plate 17 Wedding of Begum Sarbuland's daughter Usman Abida Sultan with Dr. Ilahi Bakhsh in Solan, 1931. Standing, back row, L to R: Two unknown attendants, Abdul Majeed Khwaja, Akbar Khan (husband of Qudsia Sultana), unknown British man, unknown man, Khalil Jung, Masudullah, Maqsudullah, unknown man behind Maqsudullah, unknown man. Standing, front row, L to R: British woman, Muhammad Bakhsh (father of groom), Punjabi statesman Nawab Sir Zulfiqar Ali Khan, and unknown attendant. Sitting, L to R: British woman, British man, Ajmal (Amannoo) Khwaja, the groom Dr. Ilahi Bakhsh, Muhammadullah, unknown man, and Taj Khwaja in profile. Note: Though Begum Sarbuland does not appear in this photo, she was present at the wedding, which was held at the family's ancestral summer home in Solan in 1931. The image points to Begum Sarbuland Jung's willingness to marry her children outside of her own ethnic and linguistic group (her sons-in-law Dr. Ilahi Bakhsh and Akbar Khan were Punjabi) and demonstrates her ease with the very different cultures of Delhi, Punjab, and the British colonial administration. *Source: Abid Ilahi and Taj Khwaja.*

Plate 18 A portrait of Begum Sarbuland's youngest son, Halimullah Jung. *Source: Zehra Jung, daughter of Farooq Jung.*

Plate 19 Begum Sarbuland with her nephew and son-in-law, Qulich Beg (Kuleej Baig). *Source: Saberah Malik.*

Plate 20 Begum Sarbuland with a grandson. *Source: Abid Ilahi.*

Plate 21 Begum Sarbuland Jung, possibly late 1930s. *Source: Abid Ilahi and Taj Khwaja.*

Plate 22 Begum Sarbuland's daughter Khurshid Begum, reading what appears to be a copy of *The World through a Woman's Eyes*, possibly late 1930s. *Source: Abid Ilahi and Taj Khwaja.*

Plate 23 Begum Sarbuland Jung in Nice with the former Ottoman Sultan Abdülmecid II, in 1937. To her left is her granddaughter Akhtar Khwaja. *Source: Sahar Mahjabeen Hasan and Shakira Sultana.*

Plate 24 Begum Sarbuland with family, circa 1939. Standing, L to R: Halimullah, Usman Abida Ilahi Bakhsh, Amtul Jameel (Munnoo Khwaja), Dr. Ilahi Bakhsh. Seated, L to R: Begum Sarbuland Jung and unknown European woman. *Source: Abid Ilahi and Saberah Malik.*

Plate 25 Begum Sarbuland with her eldest daughter's family. Standing, L to R: Akhtar Khwaja, Taj Khwaja, Khurshid Begum, Begum Sarbuland Jang, Sikandar Khwaja, Alima Khwaja, Amtul Jamil (Munnoo) Khwaja, and unknown woman. Seated on lawn, L to R: Unknown man reclining, Jamal Khwaja, Raveend (Rasheed) Khwaja, and Abdul Majeed Khwaja holding a young Ajmal Khwaja. *Source: Jawahar Khwaja.*

Plate 26 A family photograph with Begum Sarbuland. Standing, back row, L to R: Abdul Majeed Khwaja with hand on shoulder of Jamal (Amannoo) Khwaja, unknown man behind Amtul Jameel (Munnoo) Khwaja, Sikandar Khwaja, their two paternal cousins, husband of Abd ul-Hamid Khwaja's sister (father of Ache Miyan), Akhtar Khwaja, her cousin, Habiba Khwaja, unknown woman, Usman Abida Sultan, and Alima Khwaja. Seated on raised platform, L to R: Unknown woman in light sari, wife of Muhammadullah (daughter of Zulqadar Jung); behind them, L to R: Khurshid Begum in profile, Taj Khwaja, (reclining and in profile) mother of Abdul Majeed Khwaja, Khair un-Nisa Begum princess of Mangrol state (wife of Mahmudullah). Young boy standing: Raveend Khwaja. Seated on chair at extreme right: Begum Sarbuland Jung. Seated on floor, L to R: Unknown young girl; Azizullah (son of Masudullah); Khalil Jung (son of Mohammadullah); Sami Jung seated on lap of his father, Muhammadullah, who is holding infant daughter in other lap; Halimullah; and Maqsudullah. *Source: Jawahar Khwaja.*

Plate 27 Begum Sarbuland Jung in her summer home in Solan with her close aide and companion Bakhtawar Dulhan. *Source: Abid Ilahi.*

Plate 28 Begum Sarbuland with her grandson Farooq Jung in Solan, 1957. *Source: Zehra Jung daughter of Farooq Jung.*

Plate 29 After returning home from London, Begum Sarbuland used her old wedding dupatta to design and stitch this blouse, which she wore to the Delhi Durbar of 1911. *Source: Maryam Begum Aslam.*

Plate 30 The final resting place of Begum Sarbuland (left) and her husband, Hamidullah Khan (right), in Delhi near the shrine of Khwaja Baqi Billah in 2023. Nearby are the graves of several other family members, including that of her father, Nawab Agha Mirza. *Source: Daniel Majchrowicz.*

CHAPTER 7

EGYPT

مصر

Jeddah to Constantinople

January 5, 1910: In the morning we woke up and had tea, and then Nawab sahib went to arrange our transport. I had a little breakfast, but I'm still not feeling completely well. Meanwhile, Mahmud Asfi's wife, daughter, and daughter-in-law all came to see me. I said a few things to them in Arabic. We talked until nine in the morning. Nawab sahib came and said, "Come quickly. I've arranged for us to go out to the ship with the ship officer. If we miss this, then we will have to arrange another boat for ourselves, which may or may not be possible. The ship is anchored far away, and the wind is still gusting." I was already ready, in any case. Our things had already been sent to the port. We said goodbye to everyone and went to the port.

I'm weak from my fever and cannot walk fast, but, al-hamdulillah, we arrived at our destination on time. The ship's officer was waiting in the boat. We boarded immediately, but our baggage had been placed somewhere some distance away from us. The servants were still working on bringing us our things when the ship's officer started to make a scene, telling us to hurry. Nawab sahib said, "I'll go get them myself." He had me sit in the boat and went to go and bring the luggage. And what do you think the officer said as he left? "If you are not back in five minutes, then I am going to leave, and I'll take your wife with me. You'll be left behind." I said, "What's this? If that's your plan, then get me off the boat right now." Sarkar went quickly and got the luggage.[1]

We left Amina bi behind in Jeddah. We told her, "You travel with the caravan and bring with you whatever of our things we've left behind. Go straight back to Hyderabad Deccan with the servant Mahmud Ali. We will return later. We have arranged everything for you already." After explaining all this, al-hamdulillah, we departed.

It was very windy, and the boat moved swiftly. This is the same type of sailboat they call a bagla here. They shift the sail to match the wind's direction. It is not easy to navigate.

The sea was "rough" today, and the boat was tossed about. There are many rocks in the sea around Jeddah. They have been marked so that boats do not accidentally crash into them. The boats maneuver around them carefully. Ships are anchored so far offshore that it takes a full two hours to reach them. Our boat bounced and flounced its way through the sea. At one point it was tossed onto some rocks that made the boat shudder. Two sailors immediately leaped into the water, swam beneath the boat, and slowly freed it. I was a bit frightened. Seeing my reaction, the ship's officer, who was English, said "Don't worry, there's nothing to fear." Thank God, a thousand thanks to God, the boat was righted, and we reached the ship in no time thanks to the strong winds. We all boarded safely. This is the *Assouan*, a large, fine ship, clean and bright.[2] We were given a spacious six-bed cabin. One of the features of this ship and something that I really liked was that the women's dining table is entirely separate from the main dining area. It is marked off with large, yellow curtains. There is also a separate drawing room for women that is well-appointed too. This made me very happy.

We are traveling in first class. We arrived at lunchtime. The table was already set. We went for lunch along with the other passengers. The ship set sail at zuhr. Because I am still not healthy, I felt seasick, which I was not expecting at all. I vomited repeatedly, so forcefully that I lost my composure. Nawab sahib began to worry. He took me from the cabin and brought me up to the deck and had me lie down there. He plied me with lemon sherbets and the like until finally, at eight, I began to feel better. This nausea ended up working out in my favor because it cleared out everything in my stomach that was making me ill. At eight I had something small to eat and then returned to my cabin to sleep.

January 6, 1910: We woke up in the morning. Today I feel much better. I thanked God and got dressed. We went up to the deck to take a walk. Then we went to the dining room and had breakfast. I wrote in my diary at the table and then went on deck. There was an Egyptian officer having a walk there who was the same age as my brothers Akbar Beg and Haidar Beg. He seems very affable. He came over to me, and we began to chat. He knows a little English but speaks French and Arabic very well. After exchanging a few words, he asked me, "Do you know how to play draughts?" I told him I did, so he very eagerly retrieved his board, and we sat to play. I beat him twice. He beat me once. He was very happy when that happened. We continued playing until lunch. Nawab sahib even watched us play. Then we went to have lunch in the drawing room.

After the meal Nawab sahib told me, "Go into the ladies drawing room. There is a piano there that you can play. There are some other ladies there too." I went, but I could only remember one ghazal. Here are two of its lines:

Saqiya mai deh ki durdi-kash-e mai-khana-im
Ba kharabat ashna o ba khirad bigana-im[3]

I played it on the piano. Then a Turkish man came and sat at the piano. He played very well. The session came to an end. I am still feeling a little seasick although I am a very good "sailor." I've traveled on so many ships without ever getting sick, but this time, because of my fever, I'm not feeling well at all. Anyhow, we both wrote in our diaries until four, teatime. We had tea and went to the deck for a walk. It is cold today, and the wind is sharp. We chatted until dinnertime. When the bell rang, everyone went to their cabin to wash up and refresh their clothes. We both went back to the deck after the meal, but it was too cold, so we quickly returned to the cabin. We prayed isha and went to bed.

January 7, 1910: We prayed fajr at dawn and then washed up, got dressed, and went down to the dining room for tea. Today is Friday, so we planned to bathe and get ready for the Friday prayers. I went to my cabin to prepare my clothes. There is a women's bathing room here, and it is very good. After my bath I went upstairs to the women's lounge to write in my diary until lunch. Nawab came and wrote in his too. We all went to the dining room for lunch when the bell rang and then went to Friday prayers an hour later. In the afternoon we returned to our writing as usual. Even Nawab sahib wrote today. We passed the whole day in this way. Finally, we ate dinner, prayed, and went to bed.

January 8, 1910: We woke up and went for breakfast as usual. The ship anchored while we were eating. They said we had reached Tur, where we would have to quarantine on board the ship for three days. The idea is very unappealing, but what can I do; we have no other choice. I am on the deck now. It is 9:00 a.m. I can see that there are two other ships also quarantining. One of them has just two hours to go before they are free from this difficulty. Thanks be to God. Al-hamdulillah. There is still our ship, though, and one other besides. This situation is extremely difficult for the unfortunate people in third class of our ship, who are already ailing and suffering. They have begun to be taken off the ship with the greatest difficulty. Some of them are so afflicted by swelling that they can't even walk. Some have pneumonia. It is sad to see. The ship crew are removing the sick from the ship. They have set up a makeshift hospital on the deck I am on now. There are doctors too. They are moving patients right in front of me. There are both men and women here. Eleven of the men are very sick. They've been loaded onto boats and taken to shore. An elderly woman, a Turk, came on the deck for a while to look around, but she can't speak anything but Turkish. We wanted to speak to her but were unable to. She laughed and left.

It's nearly lunchtime. I went to the dining room and wrote a letter to [my nephew] Majid and [son] Mahmudullah while waiting for the meal. A Turkish man is playing the piano. Al-hamdulillah, Sarkar and I finished our meal and then spent the day in this way. I sewed a burqa for myself. Now it is five in the afternoon, and I am on the deck. I played draughts with the Turk that I have now gotten to know a bit.

Then it was evening. I took a short walk and then went to dinner in the dining room. Back in the cabin, Nawab sahib said, "I'm feeling a little warm today." I checked; he did have a light fever. We slept well anyhow, and by the grace of God the fever was gone by morning.

January 9, 1910: 7:00 a.m. I hadn't my breakfast yet because it was getting cold and I was starting to get a headache. Sarkar said, "It's nothing. Quickly have your meal here [in the cabin] and then go up [on deck]. You'll feel better." I ordered my meal in the cabin. I ate one egg, and that too with difficulty, before going up to the [drawing?] room. It kept getting colder and colder there, though, and finally I came down with a fever. Sarkar became worried and called for the doctor, who took my temperature and gave me some medicine, which reduced my chills. The temperature did not go down, though. This doctor is English. He is elderly, poor thing, and very kind. He also brought over a Turk Muslim doctor and asked him to examine me. He looked at my throat and examined my stomach before handing me some medicine and saying, "Only give her milk and tea." The fever lasted all evening. Finally, Sarkar brought me to the cabin at eight in the evening. The doctors checked on me. I still had a fever, so they gave me more medicine. I spent the evening groaning in pain. Sarkar looked after me with loving attention.

January [10], 1910: I still had a fever today. The doctors continued to check on me and treat me. As advised by the Turkish doctor, I was only given milk and tea. Otherwise, I didn't know what else was happening because the fever was so intense. I continued to have a fever through the night.

January 11, 1910: Both doctors came in the morning and took my temperature. They checked my stomach. I still have a light fever today. The day passed in the same way as yesterday. The doctor came again in the evening. They did not allow me to eat anything at all.

January 12, 1910: I still have a fever today, but it is much reduced. I'm very weak. Sarkar is constantly caring for me. I am being given yogurt, tea, and milk. The day and the night passed thus.

January 13, 1910: The doctors came, checked on me, and gave me a purgative. It was another unpleasant day. I also learned that an epidemic has been declared in Jeddah and that any ships coming from there are to be quarantined for ten days. Our ship was held too; today is the fifth day. All right, five more days it is. A poor man's anger only hurts himself. Yet Sarkar began to write letters and, after much effort, ultimately determined, "I will disembark in Egypt," after he learned that "the ship will be required to quarantine once more in Smyrna [Izmir]." Frightened by this news, he arranged for us to leave the ship early. We were granted permission.

At night we went to rest. My fever had decreased, and I felt good enough to fall asleep, but Sarkar continued to check on me throughout the night. He had me change clothes, and then I went to bed.

January 14, 1910: After three days of fever and no food, I have become very weak, but the doctors still will not let me have food. They did give me soup though. At least the fever is gone. I had another purgative today. The day passed.

January 15, 1910: News came today: "An order has come to put the sick back on the ship. It will depart tomorrow." The patients began to return. Most of them are from third class. The patients were packed into boats and ferried over to the ship all day long.

The ship's crew are all such good people. They helped me in every way. They brought me everything I needed, like milk and toast, right on time, and in great quantities. They would say, "Eat this, and let us know whatever else you might need. We will bring it." They all speak Turkish. One only speaks English. Two speak Arabic. I communicate with the two who speak Arabic and the one who speaks English. The night passed fitfully.

January 16, 1910: The ship weighed anchor at four in the afternoon. I learned we would reach Suez at five in the morning. Then it was evening, and I rested in bed.

In Suez

January 17, 1910: The ship reached Suez at seven in the morning. I struggled mightily to wake up, get dressed, and pack up my things. Sarkar packed some of it, and I had the stewards do the rest. I finished breakfast and got ready. At nine I went upstairs and gave our luggage to the Cook representative.

I was so weak that descending every step of the stairs was difficult for me. I got into the boat. This is a "steamboat"—that is, it runs on steam. It is called a "steamboat." Before we left, the Turkish doctor who had treated me came to meet us with a smile. I took down his address so that I could send him something or other from India. Then we set off. The boat took us to the port. The Cook man dealt with our luggage. We got down from the boat and showed our passports and so on. Our luggage was then brought into the [customs] hall and given to them for inspection. Then we went to sit in a carriage and went straight to the railway station. The train was ready to depart. God brought us here just in time. If we had come just five minutes later, we would have missed this train.

We boarded without any issues. All our luggage made it aboard too. The Suez train set off. I was hungry and weak. There was a dining car on the train. Sarkar was famished too. We walked through the train until we reached the dining car. We ordered some food. Sarkar had them make me soup, and then we went back to our carriage.

We got down at some station; I forget the name. Sarkar said, "My father once visited this station. He sat right there."[4] We sat in the same place. The train to Alexandria soon arrived. We got in the first-class carriage and had our luggage loaded. There are plenty of coolies at the stations here who are ready to help you get your things into the train, but they absolutely fleece foreigners. We got off lightly because Sarkar is experienced in these matters, and he can speak their language.

A Tour of Alexandria

We reached Alexandria at 8:30. It was evening, but the city still looked beautiful. The shops were all illuminated with electric lights. This city is in Egypt, but the capital of Egypt is Cairo. We transferred from the train to a carriage. We gave our luggage directly to the hotel representative to transfer to the hotel for us safely. The practice here is for every hotel to send a representative to meet every train and every ship and then stand there asking the passengers, "Pera Hotel? Aswan Hotel?" In this way, people select whichever hotel they like best and then give their luggage to that representative.

We handed our luggage to the man from the Aswan, which is the largest and most famous hotel in Alexandria today. The hotel's man arranged our carriage and such, and we arrived at the Aswan Hotel, which was nearby, in comfort. It is very nice. It even has a lift. Sarkar picked out a nice room, and we took the lift up to change clothes before immediately returning back downstairs for dinner. The dining room was brightly lit with electric lighting. There were hundreds of tables, each with two diners. The waiters are light-skinned and wear dinner jackets. We sat down, and the waitstaff began to bring food while always adhering to proper etiquette. Many other diners soon arrived and took their seats. There were two people at every table. I feel better now, though my sense of taste is still off. After dinner we went to the drawing room and found many newspapers available, but because I was so tired, I soon returned to the room. Sarkar came up, and, by God's grace, we had a restful night of sleep.

January 18, 1910:[5] In the morning we prayed and then went downstairs. The breakfast spread was wonderful: fruit and toast with excellent jams. We found that the butter, which was very good, was kept on ice. It is as cold here as a winter in Delhi, but they still keep the butter on ice. After breakfast Sarkar said, "You should go and sit in the drawing room. I will head to the Cook office to see about a ship to Constantinople. When I return, I'll collect you, and we will head to the museum. For now, read the newspapers."

He handed me an illustrated newspaper and left. I sat to have a leisurely read. There were some men and women coming in and out of the room, mostly European. Some would sit for a while to read. I was flipping through one of the papers when I suddenly turned a page and saw a photograph of Sarkar, the one I had had taken of him wearing Arab dress (fig. 7.1).[6] It had a complete narrative of him too. I was baffled. How had this photo ended up in this paper? I was still thinking about this when he returned. I showed it to him and asked how this could be. He replied, "I sent that photograph to Majid in Europe. He must have gotten it printed." I replied, "This newspaper circulates throughout all of Europe. Your picture will end up going everywhere."

Fig. 7.1 A photograph of Nawab Sarbuland Jang taken during his journey to the Middle East, late 1909. *Source: Daniel Majchrowicz.*

The Idols of the Kings of Egypt

I went with Sarkar in a carriage to see the museum. First Sarkar took me to see a massive pillar. Some ruler had this pillar made from stone. Its height is truly impressive. Then it was brought to Alexandria and erected. This pillar is masterful. I also saw two lions with the heads of men. Sarkar said, "These are images of two kings that have been carved into the stone, but their bodies have been made such that they appear to be seated lions." The noses on these idols have been cut off. They are ancient. Since the production of images is prohibited to Muslims, some prophet cut their noses off because "idol worship begins from images."[7] I also saw the pit from which these images and other wondrous things had been excavated and placed in the museum.

Museum

After seeing everything there, we took a phaeton to another museum.[8] There is a ticket for entry here. Sarkar purchased our tickets, and we went inside. A staff member accompanied us. This is a grand building filled with ancient objects placed in mirrored cases. These wondrous objects included carvings of such incredible quality that there is no difference between the copy and the original. Even the wrinkles and lines on their bodies can be clearly discerned. They have been adorned with clothing from that era, which is well designed. These clothes are carved directly onto the stone. You can clearly see that they have wrapped shawls around themselves or are wearing saris.[9]

Desiccated Royal Corpses [Mummies]

Well, we saw many wonders, but one thing that really stood out, and which filled me with both sorrow and astonishment, is that they used to build boxes to hold the corpses of kings. They would put money and jewels inside the box and then close it up and paint the image of the deceased king's face on the lid. These people [at the museum] have stood these boxes up, some of them closed and some of them opened, and placed them inside mirrored cases that look like thrones. The deceased are still wrapped up and preserved in shrouds [*kafan*]. Some have their faces covered; the covering has been removed from the faces of others. They have been dried out from top to bottom and are as thin and stiff as sticks.

We spent more than two hours here and saw many things. Then we went back to the hotel to eat. By then it was 2:00 p.m. Sarkar said, "Let's go to the ship. I've already settled our bill." We went to the ship and went aboard. It will leave tomorrow. We had a good sleep in a very nice cabin and had our dinner on board.

January 19, 1910: In the morning I looked around and found the ship to be very good and clean. It is called *Osmanieh* [Usmaniyya].[10] It is complete with a drawing room, a music room, a study. It is very large. Sarkar said, "There is still a long time before the ship will depart. Let's go tour the city." We disembarked and went for a tour, had lunch at the hotel, and then returned to the ship. The ship departed at 4:30. Since I am still weak, I began to vomit as soon as the ship moved. And not just me, many of the mems and angrez were sick as well.[11] Sarkar also felt nauseous but only vomited once. I was very badly affected. I ended up going to bed without eating anything and only fell asleep with the greatest difficulty.

January 20, 1910: I still feel sick. I am vomiting less but still feel nauseous. I spent the day lying down. By God's command, Sarkar is well now. The evening passed uneventfully.

January 21, 1910: I am better today. I washed up and got dressed. The mems are feeling better too. All the mems are very polite and refined. They've come from America, and two from London, just to see Constantinople. They are fond of traveling, just like us. We made friends with them all. One played the piano. One of them is with her husband, but he is a jokester. He jokes with us too. He shakes our hands and chats with us, and we also chat with him. The first three days here were very disagreeable, but today was very pleasant indeed.

In Smyrna [Izmir]

January 22, 1910: Today was also a very enjoyable day. We spent it in conversation, games, and activities. Our ship stopped in a place called Smyrna. Sarkar went to see the city. The mems and the sahibs did too, but I did not join because I am not back to my full strength yet. I learned that we would reach Constantinople at twelve, inshallah. Al-hamdulillah, the night passed uneventfully.

Notes

1. Hamidullah does not mention these interactions at all, beyond noting that the "agent promised to take me on his own boat." H. Khan, *Pilgrimage to Mecca*, 78.

2. This ship was built in 1882. It was later sold by its Austrian owners to the Egypt-based Khedivial Mail Line.

3. She sang the same ghazal on board another ship, on November 17. Her rendering of it here is not the standard form.

4. Hamidullah's father, Samiullah Khan, wrote a travelogue of his journey from Aligarh to London in 1880. Hamidullah was also on that trip. He was sixteen at the time. For Samiullah's description of the train ride from Suez to Alexandria as indicated here, see S. Khan, *Musafiran*, 42–43. This description, including its enthusiasm for Egyptian railway stations, was originally published in the *Aligarh Institute Gazette* on September 4, 1880. It was later republished in book form. S. Khan, *Safarnama-e Maulvi*, 33–34.

5. The diary gives the nineteenth, in error.

6. An image fitting this description was published in a major illustrated newspaper, *The Graphic*, on January 8, 1910 (vol. 81). The text is about three sentences long. The image and its caption are accompanied by two other images, one of a "Burmese Government Servant" and another of "A Manchu Family." These Asian images appear in a center column between articles titled "The Art of Fencing" and "The Last of the Wends: Berlin's Recruiting Ground for Maids." Thus, the newspaper seems to have turned Hamidullah into more of an exotic curiosity than a news item in his own right.

7. Though this passage seems to endorse the view that Islam prohibits images, Begum Sarbuland had a clear appreciation for all forms of art and no objection to photography.

8. In his own accounts, Hamidullah meticulously gives the full names of every place that he visits, as though he were recording a log of his movements. By contrast, his wife has little use for the names of the locations they visit but focuses rather on the points that most attracted her interest during the visit.

9. These passages underline Begum Sarbuland's lifelong interest in fashion and design.

10. Hamidullah gives *Ismailia*. Both *Ismailia* and *Osmanieh* were the names of ships in the Khedivial fleet at the time. H. Khan, *Safarnama*, 1.

11. The term *angrez* is multivalent and thus difficult to translate. It may refer to the British, to Europeans, or, in Begum Sarbuland's usage, even potentially to Christians. Here it refers to a group of travelers who have come from "London and America," but its precise usage varies throughout the diary.

CHAPTER 8

CONSTANTINOPLE

قسطنطنیه

The Sight of Constantinople

January 23: After waking up in the morning, praying, and having breakfast and so on, we packed our things and got ready. Then we went to stroll the deck and wait for the city to come into view. The city began to come into view around one. It is a large, beautiful city. It is located on the hills, such that its buildings are clearly visible.

Within thirty minutes, the ship had reached close to the city. Even after we came to a stop, none of the passengers went down [from the deck]; they were busy taking in the sight of the city. The mems asked us, "Where will you go, and which hotel will you stay?"

"In the Pera [Palace]," I replied.

"We are staying there as well."

What a coincidence.

Around 2:30 our ship began to inch almost imperceptibly forward. Nevertheless, the dock grew ever closer until we at last touched solid ground. The Turks were standing in long rows, looking out at the passengers. The hotel touts were all shouting out their hotels' names. "Pera Palace! Grand Hotel!" Like that. Every passenger also called out the names of the hotel they were to stay at and, summoning their representative, handed over their luggage. We asked the Cook representative to call for the man from the Pera Palace, which he had prebooked. These men are called "dragomen." Ours was very light-skinned and wore gold rings. He doffed his hat and salaamed us. The luggage was counted and handed to him. The entire process took so long that we were busy arranging all this until six in the evening, by which time the [street] lights had come on. The practice here is to give a tip to the ship staff. We paid everyone a tip and then disembarked. All the mems that we met disembarked with us. We all got into carriages and set off.

In Constantinople

Masha Allah, what a fine city! There are electric lights at every shop in the city. All the stores are very large. The city is lively. We soon reached the hotel. The hotel's doorway is illuminated by about forty electric lights. The door itself is very strange. It is eight-sided. You enter from one side, and as you go, all the other sides move simultaneously until you reach the interior.[1] We went inside, followed by the mems and the angrez. I sat on a chair while Sarkar went to speak with the manager.

"I will check the room to make sure it is suitable. Then I will come for you. Wait here until then."

When he left, I turned and had a conversation with the memsahib who had just come and taken the seat next to mine.

The stairs here are covered with fine carpets, as are all the landings and entry areas. There is plenty of marble. The walls are all of marble. Sarkar soon returned, having found a room to his liking. He finalized the arrangements, and we took a lift up to the room. I liked it very much. There are two beds, a beautiful carpet, a wardrobe with a mirror, a bench, chairs, a place to wash your face, and a writing desk appointed with all the necessary implements and so on. In short, the room had everything. The lighting was electric. There was also a quilt made of yellow silk and filled with the softest cotton.

We asked for all our things to be sent up, unpacked our clothes, and washed up. Then we changed clothes because it was already 8:30, time for dinner. After getting ready we took the lift downstairs. Our mem friends and the angrez came down at the same time.

Ah! Ah! Ah! The drawing hall just has to be seen. There are electric lights on the ceiling and all four walls. They are beautiful and shaped just like rose bulbs. There is a beautiful wooden floor. They are wooden because this room hosts "balls," that is, dances. There are little four-cornered tables meant for two diners each. We also went to sit at one of the tables. The practice here is that once you have selected a table, you should sit at that table every day [for the duration of your stay].

We began to eat our meal. All these light-skinned servers wear dinner suits as they serve the meal. Everything here is done perfectly. They began to bring out the food in clean, sparkling vessels. Because today is Sunday, there is a band playing.

After dinner we went to sit in the drawing room. This room is also beautifully appointed, so much so that the walls are lined with a golden satin flower print. The furniture is all upholstered with the same satin fabric. The band was playing just in front of us. Heaters keep the room toasty. Give the dial a little turn, and the heater warms up immediately.

The weather is fantastic. It is cold but not unpleasant. Half of Constantinople is in Europe, and the other half is in Asia. Between them lies the sea, which is spanned by a bridge.[2] The European portion is also ruled by the sultan, but this part of the city, Pera, where we are staying, feels just like our Secunderabad area, where the angrez make their homes. In the very same way, this area is largely populated by angrez. The name of this district is Pera.[3]

At midnight we began to feel very sleepy. Everyone retired. We got up, went upstairs, and fell asleep.

January 24, 1910: Today Sarkar left to meet an angrez friend of his who works as secretary to an [Ottoman] minister.[4] I quickly wrote some letters to Hyderabad and to my father in Lucknow. Sarkar came back at noon, and fifteen minutes later we had lunch. Then he then left again, while I went to the drawing room and chatted with my mem friends. They asked, "When will you go out [into the city]?"

"Oh, today I might just go out for some fresh air, but tomorrow I will go to the museum. It is raining today, anyhow."

"As for us, we will go to see the museum today. Where has your husband gone off to?"

"He has a meeting today with a minister and his secretary. That's where he's gone."

They left at 1:30. I went back to my room. I sat in the room for a bit and then got ready and went back downstairs to the stationery room to look through the newspapers. At 4:30 they came back again and ordered tea. They asked me to join them. When Sarkar came back, he said, "Come, let's go get some fresh air." I quickly finished my tea and went out for fresh air. We walked around for an hour before coming back.

There was no band today because it is not Sunday. The band plays twice a week. We had dinner at eight and then went to the drawing room. The angrez and the mem ladies came too. The American man who is married to one of the memsahibas is a very jovial character but also very refined. He says such funny things that you cannot help but laugh. The poor creature comes over to chat with us too. He examined my sari. His wife and sisters were with him. We all chatted for a long while before returning to our respective rooms to sleep.[5]

January 25, 1910: In the morning Sarkar said, "Come, let's get you a good cloak and coat made. We will be meeting some of my contacts here." After getting ready we went to the tearoom and had butter, toast, jelly, and jam. It was all delicious. After having tea we walked to a very impressive store. This is the city's largest and most famous shop, located in a tall building. It even has a fine marble staircase and beautiful lighted braziers. I went in and sat myself down comfortably. This shop belongs to Greek or French people. The shop owner took Sarkar's measurements and helped him select a fabric. He said, "I will tell you where to find an excellent shop for women. Go there; you can get anything you like stitched."

After finishing our business with him, we went to the ladies' shop. This shop was also very large and very good. All the workers and tailors were women. A mem came and showed me samples of many beautiful fabrics. I found one that I liked. I arranged everything and placed the order through Sarkar, who translated from French. "Come back at nine tomorrow for a fitting," she said. This was just like with Sarkar's tailor, who had asked him to come at four tomorrow.

We went into more shops after that. I bought a beautiful pair of ladies' gloves, warm pajamas, an undershirt, a warm sari, and a strong umbrella. By then it was lunch time. We returned to the hotel and ate. I went to the room, but Sarkar went out for more meetings. Once again, I spent half the day relaxing at the hotel. I went downstairs at four, where I met with all my mem friends. They asked me, "Where did you visit today?"

"I went to order a dress because I will be meeting with the sultan."

"We saw him today when his procession passed by us."

"How can I meet him in that way? He is a Muslim sultan. I will meet him in the Islamic way."

"By wearing a niqab?" they asked.

"Yes, I will wear a niqab and salaam him."

Sarkar soon returned and said, "Well, I went to meet the minister.[6] He asked me to come at three. He has promised to present me to the sultan for an audience after the Friday prayers; he has already asked for permission.

Today is Tuesday. Our clothes will be stitched by then, but I will wear Medina-style clothing to meet him on a Friday."

"Yes, of course," I replied.

Dinner was at eight, but our American mem friends and the sahib had gotten dressed up to go and eat out somewhere. After dinner we went back to the room and had an uneventful night.[7]

January 26, 1910: We got dressed and went down for breakfast. Two visiting cards came for Sarkar. He held these meetings in the room while I sat in the office room.

"You have two invitations. One is from Safi ud-Din Beg. The other is from Hamid Pasha, the secretary to the prime minister.[8] One is a dinner on Friday; the other is on Saturday. Come, let's check your measurements for the cloak and coat. I'll have to go somewhere else afterward." Sarkar took me to the shop to check the measurements, and then we came back to the hotel and ate. Then Sarkar went out again. Before leaving he said, "Today my friend Fitzmaurice will come to collect me at one thirty. I'll return by then. Once I've gotten back from the place that he is taking me to, we can go out."

"Bismillah, go ahead. I will go out again later," I said and then went to my room to sew a burqa.

It was around one fifteen or one thirty that a waiter came bearing Fitzmaurice's card. He said, "A sahib has come and is asking after your sahib."[9] I don't speak French, and this waiter spoke French but not English. Anyhow, I took the card and said, "Tell the sahib that he has gone out but will be back soon." However, he left again without understanding anything at all. A short while later there was a knock on the door. As usual, I said, "Come in." All of a sudden the sahib walks in. I was all alone. I completely forgot that this must be Fitzmaurice. I was frightened. Who could this man be coming into my room! He asked me, "Where is sahib?"

"He is out meeting someone," I replied. "Please wait, he will be back soon."

"In how long will he come?"

"He will be going out with Fitzmaurice, who is waiting downstairs. Sahib will be back at precisely 1:30."

The man took out his watch and said, "It is already 1:30. Actually, it is already ten minutes past that."

"If you are in a hurry, please leave your card with me, and I will give it to sahib."

The poor man became confused.

"*I am* Fitzmaurice, but you are speaking as though I have given my card to someone else, and now you are asking for it again."

Anyhow, he sat and waited for about ten minutes before leaving. I laughed at myself for getting scared earlier and for speaking to him as I had. I sat there chuckling to myself. Sarkar came back and changed. Then he said, "So you spoke with Fitzmaurice? The waiter had sent him up to the room. Did you offer him a place to sit?"

"I thought that they had sent someone else up. Anyhow, I did speak to him, but I mistakenly told him that Fitzmaurice was waiting downstairs and that he should give me his card so that I could pass it on to you." Sarkar laughed. I said, "If that really was Fitzmaurice, please ask him to pardon me."

Sarkar went out again, and I went downstairs at 4:30 to meet with my lady friends. I told them this story. They had a good laugh. Someone held a special dinner for one of their lady friends tonight. There was a special table that was beautifully set for the gathering, which was for about twenty people. Their table was near ours. We ate and then went to sit in the drawing room and watch the group's antics. All my friends and I gathered, and we began to chat about their dinner party. One of the mems in their party was wearing an enormous hat. We all laughed about it. That group also teased one another plenty too. After all this, I went upstairs to sleep.

January 27, 1910: I woke up, prayed, changed my clothes, and had breakfast. One mem from among our companions left today with her husband. She gave us her card and took our card before going. The American women and the man will leave tomorrow, on Friday. They asked me, "When will you leave?" I told them I would stay here for another few days. "I have dinner invitations for tomorrow and the day after. Then I will make a tour of the city and meet with people."

"Many people come to visit your husband here every day. Do you have family here?" they asked.

"Just friends," I answered.

They handed me a newspaper. "Look, he has been published in the newspaper too. They've really given him high praise."

Later Sarkar said, "Let's go walk around for a while. Then tomorrow or the day after we will go and see some sights."

"Very well," I replied, and we both went out for a long stroll before returning for lunch.

The Magnificent Mosques of Constantinople

After lunch Sarkar said, "Come, let's visit the mosques. We'll have to cross the bridge over to Constantinople proper, the area they call 'Istanbul'" (fig. 8.1).[10]

"As you suggest," I replied.

We called a carriage and got in. We had already done our vuzu. First, we visited the Sophia Mosque.[11] It is very famous and very fine. It is made of marble. The doors are inlaid with mother-of-pearl, which is well worth seeing. Costly carpets have been laid down throughout [the mosque]. The area for vuzu is very finely constructed. We prayed two rakats on entering the mosque before looking around. Then we went to the next mosque. I found it to be excellent as well. We joined with the others in praying asr here. We inspected every inch of the mosque. We saw the graves of some kings. We prayed fatihas over them and then went to another mosque. Some mosques were built by Sultan Salim, others by Sultan Mahmud, some by the sultan's mother, some by a princess, some by Sultan Muhammad. Every king has a mosque, and each is worth seeing. Today we saw fully six mosques, one after the other.[12] Each of them is unique in its own way.[13] By then it was already 7:30. Sarkar said, "Let's go to the mosque of Hazrat Ayyub.[14] His grave is there. Today is Thursday.[15] We will pray a fatiha."

"But it's already late," I objected.

"No, we should definitely go. Hazrat Ayyub was a companion of the Prophet."

At the mention of [the Prophet's] name, I immediately assented. "Let's go."

This mosque was very far, and it was drizzling, but by God's grace we arrived. We recited a bismillah and went inside the mosque. It was a very fine mosque. The congregation was ready to pray isha. We had arrived at the perfect moment. I learned there would also be a *khatm-e Qur'an* today.[16] Masha Allah, what a wonderful gathering Allah has brought us to, that the saint has invited us to. After the prayer and the khatm-e Qur'an, Sarkar and the Turks went into the shrine of Hazrat Ayyub. Meanwhile, a Turk led me to the tomb of another saint and said, "Sit here." I sat and began to read the fatiha. A full thirty minutes later, the Turks emerged again.[17] One of the Turks came and said to me, "There is now purdah inside. Please enter the shrine."[18] I went inside; Sarkar was still there. Masha Allah, what a wonderful shrine. It is beautifully decorated with fine carpets and lighted

Fig. 8.1 An artist's rendering of the view from Galata Bridge as seen in a postcard mailed from Pera in Constantinople in 1904. *Source: Daniel Majchrowicz.*

lamps. There are costly and beautiful curtains over the doors. A golden railing surrounds the shrine.

There is a well near the shrine. First, a Turk gave me water from it to drink, after which I read a fatiha. I stayed there for about ten minutes. There is also a footprint of the Prophet, which I kissed. Then I kissed the standards that were used by the army in battles and that were flown by the victors over the forts they took. They have been preserved here since the time of the Great Companions.[19] After looking carefully at everything, we returned to our carriage.[20] An elderly Turk tried to speak to Sarkar, but upon realizing that he did not know Turkish, he laughed and walked away. Everyone was very polite. We left and returned to the hotel. After dinner we said goodbye to our American friends, went to our room, and slept.

Al-hamdulillah, today, January 28, was well-spent. Because it is Friday, we had a full bath before breakfast. Sakar said, "We've been summoned at 3:00 p.m. We will take our own carriage. Please be ready by 3:00. I will be going out; the minister will collect me. You should have lunch. I may be late, so please do not wait."

"All right," I replied. "You go ahead."

He quickly got dressed in an Arab outfit because he is going to the Friday prayers. He will meet the sultan. He left, and I began to prepare my ensemble. My coat and cloak have come. I went to lunch at 1:00 on the dot. I told them that Sarkar "will come to eat any moment now." Then I went upstairs to do some reading and writing. Sarkar came at two, had lunch, and then came up to the room.[21]

"I had an excellent meeting with the sultan," he said. "We spoke for an hour, and he granted me permission to see the Prophet's mantle.[22] He said to his minister, 'Take him to see the Prophet's mantle tomorrow or on the day that he specifies. When you do, also show him the armory and the treasury—everything.' We have dinners to attend today and tomorrow, so we can go and see these places the day after tomorrow. It's nearly three now. Please get ready. How do you like your new coat?"

"It's good; it fits well."

I was ready to go by three. Sarkar also got dressed.

A Meeting with Some Turkish Sisters

The carriage came at four. We boarded it and went to the home of Safi ud-Din Beg.[23] His house is located just off a thoroughfare. The door opened, and I stepped inside. Sarkar was taken off in one direction, and two or three extremely beautiful Turkish ladies placed their hands on my arm and led me upstairs. All the houses here are multistoried. There are beautiful carpets on the staircases. When I entered the [visiting] room, I saw that it too was beautifully decorated. There were six or seven ladies there whose hair and dresses were just like those of mems. All these women were Muslim Turks. They are called *khanam* [lady]; no one knows the word *begum* [lady].[24] I asked all the khanams their names, and they asked me mine. Because I am, in fact, a khanam too, I told them my name was Akhtar Hamid Sultan khanam.[25] Their ears perked up at this. "You're from the Chukza'il family," they said.[26]

The conversation thus far was conducted partly through gestures and partly through one lady who knew some Arabic. The rest just stared for they [only] speak Turkish. We continued on in this way for an hour before calling for a translator. They asked me, "What languages do you know?"

"Hindustani and English."

They then said, "We should find a mem who speaks both English and Turkish."

One of the ladies went out and called for her husband and asked him to "please call for a translator." Thirty minutes later a memsahib came. The moment she arrived, she said, "I am very happy to have been called in your honor."

[I said to her,] "We have called you here to act as a translator. Please explain my words to these women in Turkish."

She laughed and said, "I'm sorry, but I do not speak Turkish. I speak Arabic, because I am *al-Jaish*."[27]

She did speak English but not well. Now everyone had a good laugh [at the situation]. I gestured toward the lady who spoke Arabic and said, "All right, you can translate into Arabic for her, and she will then translate it into Turkish for the rest. And I will speak to you in English."

Everyone laughed even more, and then we began our proceedings. They asked me, "How is it that your name is Sultan?" I explained my lineage to them, and they were delighted.[28] "She is a Turk too!" they said. The mem then said to me, "Until today, I have never seen a khanam before. This is the first time I have been inside the home of a Turk. It is very rare for a Turkish lady to associate with an angrez lady.[29] There are many disagreements between us."[30]

Our conversation continued during the meal. The food was varied and delicious. There is a practice here of alternating the courses between sweet and salted dishes. That is how we ate here. Although the Turks do not know how to make *pulaos* [rice-based dishes], they do make *salans* [gravy-based dishes]. The food here is good, but I didn't see a single dish that was similar to Indian food.[31] The meal continued for a long time and included a lot of fruit. They looked after me very well. After dinner we shifted to the drawing room. Safi ud-Din Beg Khanam called for her young children and asked them to sing in Turkish for me. Their ages are six, seven, five, and four. Their voices were very melodious. Then she herself played the piano, and her sister sang. They put on a bridal outfit to show me what it looks like. "This is what a bride wears here." They showed me the kind of jewelry that is placed on a bride's head; this ornament is very beautiful. Then they requested me to "please sing us a Hindustani song, if you know how to." I sang some of my own Urdu ghazals.[32]

The fun and games continued on in this way until it was eleven in the evening. Then I said, "I would like to see the old style of clothing worn by Turkish ladies, for you told me that the clothes you are wearing now are 'modern.'" The girl went to put on an outfit that was in the old style. While there was a difference between the old and the new style, much of it was very similar. Then I thanked them all and said, "I lament that I do not speak Turkish for, otherwise, on the occasion of this gathering, I should have liked to give each of my Muslim sisters-in-religion a speech."[33]

[Earlier in the night,] when I had removed my coat and cloak and given it to the ladies, they had wrapped it up in an elegant satin cloth for safekeeping. When I got up to leave, they brought [the bundle] to me. "What is inside this?" I said confusedly. One of the ladies sprung up, opened the bundle, and then herself helped me to put on the coat before wrapping the cloak around my shoulders. Similarly, all the ladies held my hand as we walked to the carriage. There was an older lady, and even she walked me to the carriage and helped me in. We said goodbye and returned to the hotel, went to our room, and fell asleep.

January 29, 1910: Today is Saturday. I washed up and had breakfast. All our earlier companions, the mems and the sahib, have left now. I've heard that there will be a ball tonight. Banners are being put up all over the hotel. They are installing even more beautiful lamps and globes. The dance will be held after dinner.[34]

An Invitation to the Home of the Secretary of the Prime Minister

Sarkar said, "Today we have an invitation to visit the home of the prime minister's secretary.[35] Please get ready just as you did yesterday." We left for his home at 4:30. This was just a visit for tea. Hamid Pasha's wife said, "You did not accept our invitation to dinner; you have only come for tea."

I replied that [hosting a full dinner] "would be a major inconvenience for you. This is already a dinner."

There were again six or seven ladies present, but [now] we had no translator. Hamid Pasha's wife speaks Arabic, though, so we muddled through with what little Arabic I could speak.[36] We conversed until 7:30. Their house is very beautiful and well decorated, large, and clean. The "compound" is nice and very large. The floors are all made of marble and overlaid with fine carpets. There are beautiful lamps in the shape of globes. These Turkish ladies are expert hosts. Hamid Pasha's wife [*bibi*] is quite old, but she looked after me very attentively all evening. Hamid Pasha's wife [*khanam*] even fed me biscuits with jam and such with her own hand.

We left at 7:30 on the same carriage that brought us here, a beautiful landau that belongs to Hamid Pasha. We arrived at the hotel right at mealtime and had dinner. The hotel had been all done up. The ball was supposed to start at ten. I went up to the room while Sarkar did some socializing. He came back at ten and said, "I won't attend the ball. I'm going to sleep."

"Shall I go?" I said.

"Yes, go ahead. You can see everything clearly from the upper floor. You can view everything comfortably from there."

Upon hearing this I quickly got dressed and left the room. After going down one floor, I could see the drawing and dining rooms directly. I took a seat at one of the windows and then noticed that many angrez men and women had gathered around the other windows to watch too; even those who were not participating in the ball had come to see it. A mem came and sat near me, but she spoke French and did not know English.

The music began, and people started to dance. There were one or two Turks participating too.[37] There were many types of dances. Another woman came to watch from my window. She did speak English. She asked if I liked the ball. "I like it very much," I replied and praised the event highly. She was pleased to hear it. I went back up to my room two hours later.

The Armory and the Treasury

January 30, 1910: I woke up and had breakfast and so on. Sarkar said, "Get ready quickly. A man has come from the sultan. He said, 'Come with me. The armory, the treasury, and the Prophet's mantle have all been opened so that you may see them.'" We quickly got ready and went to the carriage. The official came with us. At 9:30 we entered a large building.[38] The carriage stopped before a large staircase. Royal servants formed a line and salaamed us. We were led into a beautifully adorned room and asked to sit. The room was just by the sea. Next to it there was a beautiful garden decorated with the most beautiful objects. We were served coffee in cups lined with gold and jewels. Cigarettes were presented. Sarkar took one to be polite, though he does not smoke.

Then another room was opened, and we were ushered in. There we saw a royal throne covered in gold and set with diamonds, pearls, emeralds, and rubies.[39] We examined it carefully. Then they showed us another room where kings are fitted with their turbans and crowns on the occasion of their coronation. Then they took us to the place where princes are circumcised. This room is also beautifully adorned and filled with priceless objects. Both the garden and the sea were visible from here. The cases in the throne room were filled with metal dishes brimming with jewels. The number and beauty of the vessels and the other silver and gold objects are beyond expression.

Next, we went up and visited another room. This room contained paintings of all the kings of Constantinople, along with their garments. Each king wore a different type of clothing. Their turbans were also unique, but all of them were topped with diamonds and pearls set in gold. There were paintings of twenty-five kings. Each of their names was written on their chests.

Then we saw several rooms in the armory where there were many types of ancient armors and weapons, including chainmail. Many of the weapons and armor were made of gold and gold lining. Then we went to a room containing books of *fiqh* [Islamic jurisprudence] and hadith written in the hand of Sultan Salim. After that, they took us to the mosque. It is an excellent mosque and very beautiful. One room is designed in the style of Baghdad and contains the entire text of the Qur'an. There are fine carpets, drapes, and sofas. There are balconies facing in every direction and an unobstructed view of the sea. We prayed zuhr with the congregation while we were here. Sarkar tried to give the guardians two pounds, but they refused it. "We are royal servants. We have shown you this place by the sultan's decree."

Pilgrimage to the Blessed Mantle of the Prophet

After the prayer the royal servants took us to the room where the mantle is kept. This room was the most beautiful and ornate of all. Qur'anic inscriptions adorned the four walls. There is a green curtain. It is lifted from one side, and you then see that everything inside is very ornate. To one side, there was an enclosure bounded by an elaborate grilling made of gold. It was covered with curtains, which were also inscribed with Qur'anic verses. The Turks quickly lifted the curtains. This enclosure has two doors, but it is not allowed to open the second door. That door is only opened on the day when a new ruler ascends the throne. On that day alone, the new king is blessed to make a pilgrimage here. We looked inside the enclosure. There are thirty standards here that once belonged to the Companions. We prayed a fatiha and duruds for the Prophet. I stood close to the grilling for some time before coming out [from behind the curtain] again.

[The room also contained] a complete footprint of the Prophet. I pressed my eyes to it. As we were leaving, we looked at a paper with verses from the Qur'an written in the Kufic script by Hazrat Umar. It has been very carefully preserved. There are standards kept in the four corners of the room. I went inside [the enclosure] after visiting the footprint. The Blessed Mantle is kept in a silver box. I stood before it for a few minutes before leaving. The attendants would not accept anything from us. Their mastery of proper etiquette is so perfect that even if they must pass before you, they will never allow their backs to be turned to you but walk backward instead.

After visiting these sacred relics, we returned to the hotel. We had spent so long there that it was already 3:30. By the time we reached the Pera Palace, it was already past lunchtime. We just had some fruit and tea; then I went to the room and threw myself down in exhaustion. In the evening, we had dinner and went to bed.

Today is January 31, a Monday. We woke up, got dressed, and went down to breakfast. The mem that we had met at Safi ud-Din's house came to the hotel. It was 9:30. The mem said, "Come out with me. If you need to buy anything, I can help you." Sarkar said, "Go ahead, go shopping with her. Buy anything that you need, but return here by one. I will take you to the museum today." We got into a carriage. First, we went to the memsahib's house. From here Sarkar went off on his own business.[40] I stayed at her house for a bit, and then we went to the bazaar.

This bazaar is very large. This is where Turkish ladies do their buying. There are many men too. I've never seen a bazaar like this one. The shops are wonderful and very elegant. There are even jewelers' shops, lots of them. I bought some things. That mem took so long that I didn't get back to her house until four. When we arrived, I saw Sarkar's carriage standing there. I was told that he had waited for me but had finally gone off somewhere again.[41] The woman had *al-jaish* food prepared for me, which I ate. Sarkar returned in the meantime. He gave some cloth to the woman and said, "Have a good coat stitched for her." She said, "I know how to sew myself. I'll cut this cloth and come tomorrow to take the measurements." We entered the carriage and returned to the Pera Palace. In the evening we had dinner and went to bed.

February 1, 1910: I got ready and had breakfast as usual. A letter arrived from the mem. "Please come to my house yourself. I am sewing; I will take your measurements here." Sarkar said, "I'll drop you there and then go somewhere else. Then I will come and collect you again." And so I went to her house again. She is excessively talkative. As she is not English but Egyptian, she speaks Arabic well. When I looked at the coat, I saw that she had completely ruined the cut. The lining she chose was even worse. "I'm sorry," I said, "but I don't like this lining very much. It would be fine for you

Fig. 8.2 An artist's rending of the view of the European side of Constantinople from the Bosporus, as seen in a postcard mailed from the city in 1910. *Source: Daniel Majchrowicz.*

to just remove it." We went back the bazaar, where I bought a new lining and the velvet that she recommended. I also bought four gold bangles for six pounds because my wrists were empty. We ended up staying out until four again, but this time I had brought food with me from the hotel. I ate it at her house and then took a carriage back to the Pera Palace.

Today Sarkar waited and waited for me before eventually going to see the Bosporus. He returned to the Pera Palace after me. When he came back, he said, "I will take you to see the Bosporus tomorrow. Asia is located on the far side [of the strait]. That side is not very beautiful. I went there today, and it's a good thing you did not come. I had to walk a long way.[42] Tomorrow we will go to visit on the European side. It is much nicer there and makes for a better visit." I said, "Very well."

February 2, 1910: I woke up and had breakfast as usual. We boarded a carriage and left the hotel, bringing with us a packed lunch. We first went to the museum. We saw the whole thing. This museum is not very large, but it is nice. Afterward I went to the mem's place to have my measurements taken, and then I got back in the carriage and we left for the waterside. We boarded a steamboat and spent three hours going from one side to the other.[43] On the way we saw many houses and royal palaces (fig. 8.2).[44] We continued to see these things on the way back. We had our meal during the trip. We spent three hours going and three hours coming, seeing the Bosporus the whole time. We returned to the Pera Palace at seven. After dinner we went to relax.

A Meeting with the Sultan

February 3, 1910: We woke up in the morning as usual. Sarkar said, "Please prepare your luggage. I've called for a servant from the royal palace to come today. He is from Medina. He has asked me for a favor. We will go with him and see the royal palace." I quickly got ready. We had breakfast, and then at nine we got into the carriage with the royal servant and departed. When we pulled up to the palace, all the royal guards were standing at attention.[45]

The servant got down from the carriage and went inside himself first. He returned again after having received the royal orders. With him was another royal servant who said, "Please come." But then, on noticing my presence, he said, "Stop, please. I must take permission for her to enter as well." He left, then returned, and said again, "Please come."[46] We both entered the palace, along with the servant from Medina. Oh! Oh! Oh! What a wonderful garden this is, what a beautiful courtyard. Before us stood the royal palace. It started to drizzle and because we had not been given permission to carry umbrellas into the palace we scurried quickly to the veranda. We removed our shoes, and Sarkar and I stepped inside. This room is highly decorated. It has a beautiful staircase. There were large vases of cut crystal. When we climbed the staircase, we saw that there were standing on the steps several royal servants and African *khwajasaras*.[47] Using hand gestures, they told us to "go into the second room." They do not speak. I learned that "their tongues have been cut out." They are made mute so that no one may learn anything about the royal affairs from them.

We were taken to a beautifully decorated room that looked out over the sea and asked to sit. I cannot find the words to describe how beautiful this room was and how valuable the items in it were. We were again served coffee in cups lined with gold and jewels. Cigarettes were offered. I took three as a sample and tucked them away.

After an hour and a half, an African khwajasara came and said in Arabic, "The sultan will soon call for you and your wife. Please prepare yourselves." As we had only come here to see the palace, we were very surprised. But, well, "the king has ordered us to present ourselves to him, and so that is what we will do." Soon a Turkish officer came and said, "Please come." This Turk could speak English.

We presented ourselves to the sultan in a manner in keeping with the most proper etiquette.[48] The emperor indicated two chairs and asked us to sit. He then spoke in Turkish while pointing to a man. He said, "This is the translator. You speak to him, and he will explain your words to me in Turkish." We conversed in this manner for twenty minutes. He said, "I am very happy that you have come with your wife and brought her before me." The sultan also speaks Persian. The first time they met, he spoke to Sarkar in Persian. Then he said, "Please accept a memento from me." Then Sarkar spoke in English at great length and with full adherence to proper etiquette, in praise of the sultan. After a long conversation, the sultan stood up.[49] As soon as the king arose, the translator, Sarkar, and I also stood up. We bid him farewell with great reverence while walking backward to the door. Then we walked through the entire palace, seeing everything as we made our way to the exit. I am unable to describe the adornments of this palace. It can only be seen.[50]

We left the palace grounds with the royal servant. I am very happy to have had the honor of meeting the sultan today. We were in very high spirits as we got into the carriage with the royal servant. Another royal servant who carried a tasbih[51] in his hand then came up to the carriage but only spoke through gestures. I asked, "Why doesn't he speak?" Sarkar said, "Didn't you notice in the palace that all the servants that we saw walking about, including those who served the coffee, were mute? Their tongues were cut out when they were children or something like that so that they cannot speak and thus can never reveal any royal matter. The African man is a khwajasara, and he is permitted to go in and out of the palace." That's when I realized.

The palace servants helped us into the carriage and then walked alongside us for quite a distance before giving a salaam and departing. When the carriage got under way, we saw that the sultan was just departing from another gate along with a procession of carriages for women. As a result, it felt like our own carriage had joined the royal procession. Their procession went on and on, one carriage after the other. Traffic [on the main road] was stopped for us too, since we were exiting from the royal palace as well. For this reason, we had quite a bit of fun during the ride. I was laughing the entire time at the idea that we were surrounded by officials, with royal officers in front of and behind us. Our carriage and the sultan's traveled together for twenty minutes before he went one way and we went another.

The Palace of Abdülhamid Khan[52]

We then went to into the palace of Abdülhamid Khan to have a look. First, we drove through the entire grounds in the carriage. We visited the museum and saw the bodies of many rare animals, which had been stuffed with dried grass. Then we got down from the carriage and entered the palace, which stretches out for miles and miles. There is no way that I could describe the beauty of its decorations or of its garden. Everything is beset with pearls and gold worth hundreds of thousands of rupees. Every table, every chair, is a wonder unto itself. Thousands of species of valuable animals were being raised in enclosures. Every type of animal is there, from four-legged creatures to winged creatures. There were colorful animals the likes of which I had never seen before. One particularly wonderful thing I saw here was an orange and yellow parrot that I thought very beautiful. It speaks Turkish and is very talkative.

We continued our tour of the palace from two until six, but the place is endless. We finally left because we were tired. This "palace" has been built on the top of a large, lush hill. They have a ticketing system these days. Anyone who wants to see it can buy a ticket and go in. The ticket costs one franc. At six we returned to the hotel with the royal attendant. Sarkar had him dropped off on the way as his home was nearby. Before that, though, he took Sarkar to visit an aged saint. I stayed in the carriage because I was tired. He is a great saint.[53] He gave Sarkar a book and granted him an ijazat for the Litany of the Sea.[54] This saint is a master of the Litany of the Sea.

After dropping off the Medinan servant, we went to the hotel, washed up, and got ready for dinner. We went downstairs and milled about the drawing room for a while before taking our seats at the table. Then I went to the room. I learned that there is another ball today. I watched the ball and then had a very good night's rest once again.

A Gift from the Sultan

February 4, 1910:[55] Sarkar went to meet with someone after breakfast while I started packing up my things. We leave tomorrow. Sarkar returned for lunch but then went out again, saying, "We're leaving tomorrow. Please prepare."[56] I spent the afternoon in my room. Sarkar came back at six and said, "The sultan's man has come. He is calling for us. The sultan has sent us something." When I heard this, I got up, put on a Turkish burqa, and went downstairs with Sarkar.

The sultan's man was standing in the drawing room wearing a very fine Turkish coat, holding a bag in his hands. We first asked this royal messenger to sit. He sat down. Everyone in the hotel was watching us. First, he said a prayer for the sultan. Then he opened the bag and placed the blessed gifts in our hands. We both stood and pressed them to our foreheads and eyes. They spoke in French for a few minutes. We returned upstairs after the messenger had left. We each opened our respective gifts, which were wrapped in silk and bound with silk ribbon. I wanted badly to tear it open and see what was inside. Sarkar was given a cigarette box with exquisite metalwork worth two thousand rupees. I was bestowed with a gold watch beset with jewels and hands made of diamond. It must be worth a thousand rupees. Ah, there was also a mouthpiece for a hookah inside the cigarette box. Everyone in the Pera Palace is talking about how we have been honored.

Departure from Constantinople

Today is the fifth of February, a Saturday. After breakfast, we were ready to leave. Sarkar went out on some business but returned at lunchtime. After the meal he said, "Let's go. Everything has been arranged for our voyage." We said goodbye to everyone at the hotel, took a carriage to the port, and boarded the ship.

We found the ship in a filthy state. Sarkar was very angry. "The Cook representative tricked us. He said this was a wonderful ship when he sold me the ticket. I should have inspected it first." We went to our cabin, which was very large but still very dirty. The ship departed at 7:00 p.m. Until then we stayed on deck watching the goings-on and looking out at Constantinople. I was sad to leave this city. I could not forget the sultan's kindness toward us, nor will I ever.

We and the one or two other passengers had our dinner. At least the food is good. After the meal we went to sleep.

February 6, 1910: In the morning after breakfast, we saw that the ship had stopped at an island. We departed again after an hour. Soon after the departure we encountered "rough" seas.[57] I felt nauseous. Sarkar tried to make me feel better. We played draughts. He did everything he could, but the seasickness persisted. I was barely able to have my meal. I lay down until evening and then finally returned to the cabin and got into bed. The seas were so rough today that even Sarkar felt unwell. And then I vomited. Sarkar retired early. I also went to sleep without eating anything. The night passed.

Notes

1. A revolving door.

2. In fact, the first bridge in Istanbul to span the Bosporus (that is, spanning Asia and Europe) was not built until 1973. She is surely referring here to the Galata Bridge, which spans the Golden Horn, a waterway that divides the European side of the city in two.

3. The area is better known today as Beyoğlu. It had a large European population and was significantly more developed than other neighborhoods in Istanbul. Thus does Begum Sarbuland compare it to Secunderabad, which was founded as a British cantonment and housed a significant British population prior to Indian independence.

4. In fact, he first visited the British embassy, where, armed with an introductory letter from Lord Cromer, he went to meet the ambassador.

5. Hamidullah records meeting with several Members of Parliament who were staying at the hotel.

6. He had gone to meet the grand vizier (*sadr-e azam*), Ibrahim Hakki Pasha (1862–1918).

7. Hamidullah found the evening somewhat more eventful: "In the evening we toured various parts of Pera and were introduced to many important people in the hotel." H. Khan, *Safarnama-e Qustuntuniya*, 36.

8. She uses the term *madar ul-maham*. There was no such position in the Ottoman administration, and Begum Sarbuland seems to borrow the term from Hyderabad's administrative structure, where *madar ul-maham* referred to a prime minister. Hamidullah refers to him as the minister of waqfs.

9. This was probably Sir Maurice Fitzmaurice (1861–1924), a civil engineer who had worked in the Ottoman Empire. He is described by Hamidullah as a "subordinate of the ambassador." Fitzmaurice gave him a tour of the embassy and promised to come introduce him to leading figures in Constantinople. H. Khan, *Safarnama-e Qustuntuniya*, 34.

10. That is, they crossed the Golden Horn via the Galata Bridge.

11. The Hagia Sophia, or Aya Sophia, mosque was originally a Byzantine church. It became a mosque in 1453 with the Ottoman conquest of Constantinople. In 1935 it was converted to a museum and then remade into a mosque again in 2020. When Begum Sarbuland visited, it was a mosque. The famous mosaics in the building were not uncovered until 1930, thus she would not have been able to visit them. Hamidullah calls it the "Abba Sofia." H. Khan, *Safarnama-e Qustuntuniya*, 37–42.

12. The mosques they visited are detailed in Hamidullah's account. H. Khan, *Safarnama-e Qustuntuniya*, 37–48.

13. Her enthusiasm for visiting numerous mosques contrasts with her entries from Italy, where she quickly tires of visiting touristic sites.

14. The mosque located near the shrine of Abu Ayyub al-Ansari, located in the eponymous Eyüp District of modern Istanbul. Ayyub al-Ansari was a close companion of the Prophet. He died while on a military campaign in Constantinople. His shrine is a major pilgrimage site.

15. Shrine visitations are typically performed on Thursday evening. By Islamic reckoning, days begin at sunset. Thus, "Thursday evening" is actually already Friday, the holiest day of the Islamic week.

16. A full reading of the entire Qur'an or the final session of such a reading.

17. The men had perhaps performed a Sufi zikr inside the shrine during this time.

18. The shrine had been vacated of all men barring her husband so that she could enter in a state of purdah.

19. The sahaba, the companions of the Prophet.

20. In his account, Hamidullah does not describe their visit at all but only reviews the history and physical description of the mosque. H. Khan, *Safarnama-e Qustuntuniya*, 42–43.

21. Hamidullah's description of the morning is lengthy and exuberant. As it was Friday, the sultan went on procession to his Friday prayers, accompanied by large crowds and great

fanfare. Begum Sarbuland seems to have missed out on this spectacle entirely, perhaps because the day's events were largely limited to male participation. As with his other engagements in Istanbul, his participation in the festivities was coordinated via the British embassy, which appointed him an assistant as a guide. He and the assistant went to the mosque selected for the sultan's weekly prayers, where they met him. He met and had coffee with the imam and the sultan's "special assistant" before joining in the prayers. H. Khan, *Safarnama-e Qustuntuniya*, 49–51.

22. The Sacred Mantle of the Prophet (Urdu: Khirqa Sharif) is a cloak once owned by the Prophet. It was transferred to Istanbul by Selim I in 1595. Hamidullah claims in his travel account to have had an audience with the sultan "for about twenty minutes." They conversed variously in Persian and via a Turkish translator. He spoke with the sultan's assistants in French, English, and Persian. H. Khan, *Safarnama-e Qustuntuniya*, 52–53.

23. They were joined by "Ismail Kemal Bey, member of Parliament, who was also staying at the Pera Palace." This was likely Ismail Qemali (1844–1919), an Albanian statesman who is regarded as the founder of modern Albania and its first prime minister.

24. Both words have equivalent meanings in Turkish and Urdu. The word *begum* was itself the female form of an old Turkish word for nobility, *beg*. The male form endures in contemporary Turkish as *bey*, but clearly the feminine form was unfamiliar to the Ottomans, who referred to Turkish noblewomen as *hanım* (spelled *khanam* in Urdu), from the female form of the word *khan*.

25. This compound name seems to have been composed of several elements: her first name, Akhtar un-Nisa; Hamid, her husband's name; Sultan, a title indicating royal descent; and khanam, indicating her status as a noblewoman.

26. Chaghtai, presumably. The Mughal family, of which she was a descendent, was in fact Chaghtai Turkish.

27. The reference is unclear, though *al-jaish* means "army." As is soon revealed, this "mem" is in fact an Egyptian Christian. She may have belonged to a military family. Remarkably, despite meeting with this woman several times, Begum Sarbuland never gives her name but only refers to her as "the memsahib."

28. She presumably explained that she was descended from a Turkic line. She was a descendent of the Mughal emperor Shah Alam II, who was himself a direct descendent of Babur, and through him, to Amir Timur, a renowned fourteenth-century Chaghtai Turkic chieftain. It was this connection that led the ladies to declare her a Turk.

29. As previously noted, the term *angrez* has many valences in Begum Sarbuland's diary. The woman is ultimately revealed to be an Egyptian Christian, though the issue is muddled by Begum Sarbuland's tendency to see religious affiliation in ethnic and racial terms. According to the quotation in the diary, the lady here referred to herself an "*angrez laidi*," which seems unlikely.

30. This conversation is recorded quite differently by Hamidullah, who was, of course, not present: "Begum sahiba was invited to the women's dinner, and she left. Because Begum sahiba (Mrs. Sarbuland Jung) descends from the [Turkic] Mughal family, the Turkish women considered her one of their own and treated her with the greatest love and affection. . . . The Turkish women chatted with Begum sahiba in Arabic, but one of the guests was a Christian mem sahiba who was able to speak English and Arabic well. But since Begum sahiba already knows Arabic, and also speaks English, the memsahiba was not much help. Everyone got by simply by speaking Arabic." Her account emphasizes her challenges to communicate with the women; his account emphasizes their easy unity. H. Khan, *Safarnama-e Qustuntuniya*, 51–52.

31. Recall that Begum Sarbuland almost exclusively took her meals at the European Pera Palace hotel. This was thus probably her first, and perhaps only, dinner in the Turkish style.

32. Begum Sarbuland's descendants recall that she was an avid poet, though not a remarkable one. In his account of the evening, Hamidullah writes that she recited ghazals that both

she and her father had written and that "the Turkish ladies were delighted by hearing the Indian language." H. Khan, *Safarnama-e Qustuntuniya*, 52.

33. For an analysis of Begum Sarbuland's description of this interaction, see Majchrowicz, *World in Words*, 167–70.

34. Hamidullah remembers the day somewhat differently: "Today we spent the morning touring and seeing the sights."

35. Begum Sarbuland refers to him as "Hamid Pasha" and the secretary to the prime minister. By contrast, her husband says this was "Hammad Pasha, the minister of charitable trusts [*auqaf*]." H. Khan, *Safarnama-e Qustuntuniya*, 54.

36. Hamidullah had more confidence in his wife's abilities: "The women in the zenana got along by speaking Arabic." H. Khan, 54.

37. Hamidullah records that this was an Armenian ball. His account suggests that they returned early to watch it and that he himself attended. H. Khan, 54.

38. They went to the Topkapı Palace, which would not become a museum open to the public until 1924. Hamidullah calls it "the old palace," a reflection of the fact that the Ottoman sultans no longer lived there but in the European-style Dolmabahçe Palace located several miles away. H. Khan, 55.

39. "Here we saw a beautiful throne made of gold and covered with jewels. Selim I took it from Iran after defeating Shah Ismail in battle in 1514." H. Khan, 56.

40. In his travelogue, Hamidullah writes that on this day "we" went to visit the court and a new university, though as her own diary makes clear, she did not herself attend. His use of the word "we" to refer to both himself and he and his wife thus produces significant ambiguity in his text about his wife's movements in the city that is only clarified by access to her telling of the events. H. Khan, 64.

41. Oddly, Hamidullah also seems to have spent the afternoon at the bazaar, though they apparently visited it separately. H. Khan, 66.

42. He visited a large number of sites on the Asian side of the city and committed several pages of his travelogue to describing them. He did seem somewhat unimpressed but also likely had little motivation to retrace his steps with her. H. Khan, 66–71.

43. Steamer tours would typically move up the strait in a zigzag pattern, stopping briefly at sites on either side.

44. They made a standard tour of the Bosporus, boarding the ship near the Galata Bridge and heading northward. Begum Sarbuland mercifully elides a description of the sites they saw on this tour, but they are covered in detail in Hamidullah's account. H. Khan, *Safarnama-e Qustuntuniya*, 71–78.

45. This was Dolmabahçe Palace, the newer, European-style palace and the primary residence of the sultan.

46. Hamidullah does not mention this momentary delay as permission was sought for his wife's entry. H. Khan, *Safarnama-e Qustuntuniya*, 79.

47. Khwajasaras are court eunuchs. For a study of African khwajasaras and their role in the Ottoman court, see Junne, *Black Eunuchs*.

48. Mehmed V (r. 1909–18), the penultimate sultan of the Ottoman Empire.

49. Hamidullah states that they conversed for twenty minutes but gives no indication of what they discussed. H. Khan, *Safarnama-e Qustuntuniya*, 80.

50. Hamidullah nevertheless gave it his best shot. H. Khan, 80–81.

51. A kind of Muslim rosary.

52. In Begum Sarbuland's book, the name is given as Abdülhamid Palace. Located on a hill in Beşiktaş, the Yıldız Palace was a complex of villas and plazas, which had once served as a residence for Sultan Abdülhamid II, who felt that his palace at Dolmabahçe was vulnerable to attack by sea. As Begum Sarbuland describes, it had by then been converted into a park.

53. Hamidullah describes him as belonging to the Shadhili Sufi order and the leader of a hospice.

54. Hizb al-Bahr in Arabic. An ijazat is a certification of mastery. Hamidullah does not mention this particular litany by name.

55. The text reads "January."

56. He had gone to visit a Mevlavi Sufi hospice, where he was given an ijazat for Rumi's *Masnavi*. H. Khan, *Safarnama-e Qustuntuniya*, 83.

57. "Ruf samundar."

CHAPTER 9

EUROPE

یورپ

In Athens

February 7, 1910: I awoke in the morning to discover that the ship was moving very slowly. I washed my face and had tea but still did not feel well on account of having vomited last night. The ship soon stopped. We were to disembark here and then take the train for an hour.[1] This would bring us to Athens. Our bags were already packed. Sarkar left them with the Cook representative. We disembarked from the ship and boarded a carriage. Here carriages are available everywhere, at every station. We took this carriage to the railway station. I went to sit in the waiting room while Sarkar left for a short time to see to our luggage and purchase train tickets. We boarded the train. After an hour we were in Athens. The train stopped, and we disembarked and entrusted our luggage to the Cook man. There is a Cook agent at the [main] train station in every city.

We got into a carriage and went to the Hotel Angleterre.[2] This is a nice hotel, but it is not so vibrant as the Pera Palace. We put our luggage in the room and washed up for lunch, which was already prepared. Then we ordered a carriage and went to see the famous things. We saw some famous old buildings and hills. We saw a beautiful church. We toured around and then returned to the hotel. The city is beautiful, but the people are not good in their dealings. They are dishonest.

Socrates's Prison

February 8, 1910: Today after breakfast we went out in a carriage. The place we went to today was a tall mountain with a large temple on it that even now attests to the greatness of the distant past.[3] The wonderful objects that have been found in this temple have been placed in a set of secure rooms that comprise a sort of museum. There are awe-inspiring idols sculpted from marble that are very worth seeing. This temple and everything found in it have been put on display after intensive excavation several yards beneath the surface. They have discovered large rooms and chambers whose floors are made of marble. The door frames, walls, and pillars, all of marble, are exceptionally beautiful and very tall. The marble ceiling has been made with remarkable artistry. But—Allahu akbar!—this temple has been built on such a steep mountain that I was exhausted just walking up to it. It is twice or thrice as high as Baba Sharaf ud-Din's hill.[4] The area at the top is expansive. We looked at everything, including the museum. Everything about this place, including its dates and history, are recorded in a red book that Sarkar always carries with him and from which he reads aloud to me wherever we visit.[5] There were many other travelers here who had come to see it too. They were also touring the site.

We continued our visit for a few hours before descending again. Then we got into a carriage, and Sarkar directed us to the place where Socrates used to debate. This is a plaza; there are some ancient ruins, but it is otherwise very clean. Then Sarkar brought us to where they imprisoned Socrates. It is a subterranean pit in a mountain with an opening for ventilation. We went inside and had a good look at this place too. Sarkar has a deep love for Socrates. He is almost brought to tears when he reads the history of this man. As a result of Sarkar's love for him, I also expressed my own deep sympathy for Socrates.

For lunch we went back to the hotel before resuming our tour in the carriage again and departing for another site. Allahu akbar! Was that a road we traveled on or Satan's own intestines? After an hour and a half, we arrived at our next destination. Again, we saw yet another old temple that had been dug out from the earth. There was more marble and, once again, a museum displaying excavated objects. The excavations are ongoing. New objects are discovered continuously. There were more large rooms, courtyards, halls, bath halls for the Greek gods, and more. All of this has been excavated. These gods also had children. We continued touring the site until five. We also saw a church. Then we got back into the carriage and traversed that same long path back to the hotel for dinner. Then we rested.

The Trip to Corinth[6]

February 9, 1910: After breakfast Sarkar said, "Come, there is a railway [station] not far from here. We will go from there to Corinth. We will stay there for a day to see the sights, and then we'll go to Rome and Paris." We went to the station around ten o'clock and left on an 11:00 a.m. train, arriving at Corinth at 3:00. We had lunch in the hotel's "refreshment room," which was just front of the hotel, before checking into our room. By then it was four. Sarkar and I went out for a walk and some fresh air. The sea was just in front of us. We walked a long way. This town is very small, like a village, but this country is so rich that even the villages have hotels and beautiful buildings. We went to see the church. Even such a small place has its own large, beautiful church. We returned to the hotel at 6:00 and had dinner at 7:30. I wrote a letter to Khurshid. The hotel owner sat with us for a while, conversing with Sarkar in French and showing him pictures of various local tourist sites. Then we went to our room to rest.

February 10, 1910: After breakfast we got into a carriage and set off. Half an hour later the carriage climbed a hill. Beyond it was an even higher hill that was ascended by a walking path. At the top was yet another old temple. I was not very interested in it. Sarkar read aloud to me and explained many things. I decided not to walk up; Sarkar didn't go either. We quickly left this place and went to visit the other side of the city. Here the mountains have been cut down the middle, creating a pleasant lake that ships may sail through.[7] People come from afar to see it. They use wide vessels shaped like slabs of wood to ferry people, horses, cows, and everything else from here to there and there to here. We got down from our carriage and took one of these boats over to the other side, which is not distant at all. While we were on the boat, we saw that a ship was coming down the canal. When we reached the other side, we waited there so that we could see the ship pass from up close. The boatsmen quickly moved their vessels out of the way. There is a chain that runs across the canal. They loosened it and lowered it down into the water. After about ten minutes the ship passed through the lake and the hills on either side, moving swiftly. I watched it for a long time. Sarkar wandered off quite a distance to look at everything. I collected a few pebbles from here. We took the same boat back to the other side and got into our carriage to return to the hotel for lunch. Then we went quickly to the railway station. The train pulled into the station. We departed at three.

In Patras[8]

We arrived at Patras at eight in the evening. The Cook employee was waiting. We gave him our luggage and found a carriage by ourselves. We went to the hotel. Dinner was already ready. We ate and went to rest.

February 11: After breakfast we took a carriage to explore the town from one end to the other. We saw a church in the center of a garden. The local jail is in an old temple that is very ancient. We saw it and then slowly made our way back to the hotel. I didn't go out again. I wrote some letters and so on. I learned that the ship to Brindisi will arrive at nine tonight. In the evening we had dinner and waited for the ship to arrive. At nine we were told that "the ship is late. It will be delayed."

Departure from Patras

We waited until the ship arrived at 10:30. There was a steady rain and frequent lightning. The conditions being such, everyone opened up their umbrellas and made a run for the ship. We did not need a carriage because our hotel was located right by the coast. We boarded a boat. The boat had no covering. It was a dark and stormy night. The boat set out. It took fifteen minutes to reach the ship. We went aboard along with two or three other angrez and some mems. This ship belongs to the Robotino Company. It is very nice and clean. Our luggage arrived soon after. The ship is brightly lit with electric lights. There are about ten English ladies aboard who, like us, are on a world tour. I went to the cabin. It is very nice. We inspected our baggage and then dismissed the Cook representative. After changing our clothes and so on, we slept.

February 12: I woke up, arranged my clothes and other things, and went up on deck. I do feel nauseous on this ship but only a little. We stopped somewhere at eight, and all the ladies, as well as Sarkar, disembarked to see the city. He asked me to join, but I said, "I'm tired. You go ahead. I'll stay on board." They returned an hour and a half later. Sarkar bought ten pieces of wooden carvings here because they were well made. We set off again. I had lunch and then lay down so that I would not feel nauseous. We stopped at four, loaded some cargo, and then set off again an hour later. I don't feel very nauseous on this ship because of all these quick stops. I got to know the ladies on the ship well. Each of them gave me their card, and they noted down my address. We went to rest after dinner.

Arrival in Brindisi

February 13, 1910: We arrived at Brindisi at five in the morning. We quickly woke up, got dressed, had breakfast, and then gave our luggage to the Cook man.[9] The ship docked at the port, so we easily disembarked and took a carriage directly to the railway station. This is a nice city too. I saw as much as could be seen from the window of a carriage.

A Journey to Rome

We boarded a train to Rome at 7:00 a.m. The ladies from before are traveling in the carriage next to ours, but they are going to Naples, and we are going to Rome, so their carriage will be removed from the train at four and sent in another direction. They will be our fellow travelers until four, though. They will come to Rome in five days, but when they come, we will be on our way to Florence. For lunch we had food we bought at the station. They have an excellent system here whereby they give you an entire lunch in paper packaging along with a bottle of water and paper cups. And this is how we were given our lunches. Sarkar bought the meal, and we ate and drank our fill. I spent most of my time with the ladies. When we reached the junction, I returned to my carriage, but I was with them until four, when their carriage was removed and sent on its own way, while we went our own way. We reached Rome at nine. We gave our luggage to the Cook man and went to the Hotel Quirinale.[10] The luggage arrived. We had dinner and slept.

A Tour of Rome

Today is the fourteenth of February. Although this hotel is exceptionally grand, the manager and the staff are all very dishonest, bad people. They do not tell you how things work, and then they ask you to pay them a franc and two shillings for every little thing. In a word, dishonest. This is the only time I have ever felt frightened in a hotel.

Anyhow, we engaged a carriage and took it to the museum. This city is indeed very beautiful and clean, and we enjoyed the views on the way to the museum. There were many European men and women here. I was wearing a Turkish burqa, which caused them all to stare at me and ask, "Well, who is this?" We bought tickets, went inside, and looked at the wondrous objects in every single room. We spent two hours there. The museum feels endless. I did not appreciate the friezes [or sculptures] very much, but I did really enjoy the picture gallery. It holds priceless paintings, the likes of which I had never seen before. We kept looking around until one in the afternoon. Then we returned to the hotel for lunch. I was tired afterward, so I went to the room to rest. Sarkar went out again and returned at maghrib. We went downstairs at dinnertime. Then Sarkar went to the drawing room to read the newspapers. I went upstairs. He came up thirty minutes later. By God's command, the night passed uneventfully.

February 15, 1910: We woke up, washed up, got dressed, and called for tea to be sent. We had a very leisurely breakfast, and then Sarkar asked me to get ready. "Please hurry, we are going to see the most famous church [in Rome]." We took tram number 14 to the church because this tram goes there directly. There were many ladies and sahibs on the train. The church is indeed very beautiful, ornate, and large. I have seen many churches here and found them all to be almost pitch-black inside. This one has some light inside, but it was nevertheless a light not worthy of the name. The building and the decor are excellent. There are no carpets on the ground, as with our mosques. Instead they have fine chairs, tables, paintings, harmoniums, and *botachios*.[11] The walls and ceilings are covered in gold work. The building itself is also very nice. We had a nice look around before moving on.

From here Sarkar took me to a garden that was right in front of the church but located on a slope. The streets are very clean. Sarkar asked, "Should I find a carriage, or will you walk?" I decided to walk, so we both climbed the hill together. The further we walked, the more beautiful the garden became, and slowly the entire city of Rome came into view. When, after much climbing, we reached the top, we saw that there were thousands of people there, with more constantly arriving. This is a popular visiting spot. Every morning and every evening, people come here to take in the air, children and adults alike. We also had a good walk around before descending from the park on the far side.

We hired a carriage to take us to the hotel. It was 5:00 p.m. I went up to the room but forgot to collect my key from downstairs. Sarkar was still downstairs. I went into another room and rang the bell. A maid came, and I was just asking her to retrieve the key for me when a memsahiba whose room was near mine came over. I mistakenly thought that she was some member of the hotel staff. I turned away from the maid and was just about to ask her to help me collect my key when I realized that, actually, she was a lady. I apologized immediately, but she was very kind and spoke to the maid and insisted on helping me retrieve the key. She also invited me to her room: "Until your room is opened, please sit in mine."

Then, what else, but we began to chat and in fact continued to chat even after my key was brought. The poor thing is so sweet that she has decided to consider me her sister. By chance, Sarkar was going to attend the opera tonight. I said to her, "Why don't you come and join me in my room after dinner? My husband is going to be away at the opera." She agreed, and after dinner Sarkar left, and the memsahib and I went upstairs together. She sat in my room, and I brought out my sewing. The poor thing took up the other end [of the garment] and began to sew as well. She began to tell me about herself. "I had been married for twenty-one days when my husband fell from his horse and died. He was an army officer. I receive fifty pounds a year as a pension. My name is Widow MacDonald." She showed me the various medals her husband had received. He was a well-known man. She showed me pictures of her mother and her husband. We continued our conversation until midnight. When Sarkar returned, she departed after showing me her affection. She is about thirty-five. The poor thing is spending her life honorably. May God help her. After seeing the memsahib off, we went to bed.

February 16, 1910: We woke up, prayed, had breakfast, got dressed, and then took a carriage to the most famous churches. We reached the first after twenty minutes and went inside. The interior is round and has an open ceiling that makes it very bright. This lets you see the building's design clearly. We visited this church for thirty minutes before returning to the carriage. Then we went to see two royal palaces that were located on tall hills. They were so large you must walk miles and miles to see them fully. History records that two boys named Roma were found on these hills and raised by a she-wolf who fed them with her own milk.[12] We bought two tickets and went inside. We were joined by an angrez who spoke French.[13] He showed us all the rooms and all of Rome, which is visible from here. These buildings are now in ruins, but their high ceilings still attest that this place must once have been in a class of its own. We continued slowly up the hill, carefully inspecting every item and viewing every site of interest along the way, until we climbed a staircase and entered a vast grassy plain. How beautiful and large it was! There was a wonderful garden here too. We strolled through it, chatting with the angrez, until we reached a place where a man and a woman stood putting color to a painting they were making. We looked and saw that they were making a painting of a stone statue that was leaning against a tree before them. By God, their rendition of it was perfect. I watched them for three or four minutes. I turned back only when Sarkar called out to me.

From here we had to descend the hill until we reached a courtyard and rooms with large floor mosaics. The walls were covered with ancient images painted by hand. To one side, there were three rooms, all right next to one another, where the emperor's daughter once lived. We had a good look before returning up the hill. There was a constant flow of visitors here. Then we went to another area. The angrez continued to point out and explain every site, while Sarkar read the history of the place aloud [from his guidebook]. We continued our ascent until we could see the whole of Rome. We could see outlines of all the large buildings and halls. We saw the emperor's bathhouse. At another place, a man came and, with a large broom, quickly swept away the dust to reveal a vast and beautiful floor mosaic. He said, "Once upon a time, the entire floor was like this," and then he covered it up again. We saw so many things, one after the other, until well after lunch time. We walked back down at four, visiting more places as we went. We saw where the emperor would sit. At long last we reached the place where the wolf once lived and where the two Roma boys were found. This place is inside the mountain and is something like a lion's den. It is empty now. We continued to visit all these places until we reached the bottom. I looked at my watch; it was 4:30. I was exhausted after five hours of walking.

Sarkar was tired too, but he wanted us to visit one last site, the place where the ancient and famous Roman bazaar was once held. The floor was marble. The shops and the walls were all marble. There were the famous arches. The sun was shining, and everything could be seen clearly. We saw the place where goats are butchered. By then I was completely exhausted. I was struggling to make my way back when suddenly, while we were passing through the bazaar to find a carriage, one or two hundred priests wearing long black overcoats appeared. They were processing through the bazaar right before our eyes. We followed behind them, wondering where they could be going.

There were a few people here making paintings and drawings of the ruins of this bazaar. We walked over to where the priests were gathering. We saw that there was an old church inside the bazaar with handmade paintings of Mary and Jesus, the Spirit of God, hanging on the walls. The paintings were covered by curtains so that what was left of them could not be further damaged. The priests had come to see these paintings. There was a shallow pool inside the church too.

We then left this place, exited the bazaar, and hired a carriage back to the hotel. We arrived at six. I went up to the room. Sarkar stayed downstairs to take care of some business. I washed up, changed my clothes, and went to Mrs. MacDonald's room. She spoke to me affectionately. When she learned how tired I was and how sore my feet were, the poor thing knelt and began to massage them herself. I implored her not to, but she had her maid bring hot water anyhow. She massaged almond oil into my feet and then had me soak them in the hot water before vigorously drying them off and putting socks on me. "Come with me to the market tomorrow," she said. "I'll make sure to find you warm socks and a good petticoat. I'll ask your husband to make sure you're safe from the cold."

We were going downstairs to eat together when we bumped into Sarkar. I introduced him to Mrs. MacDonald. She asked him about purchasing me new clothes right there in the hallway. I said, "All right, I will go with you tomorrow to get new clothes. Sarkar will not forbid me. I already have many warm clothes with me. The only things I need are a petticoat and an undershirt."

In the dining room Sarkar and I went to our table, and she went to hers. After dinner she said, "A friend of mine would like to meet you. Please do come downstairs to meet her." Since Mrs. MacDonald speaks with a very strong English accent, I thought she had said, "My friend and I will come up from downstairs." I thus replied, "Very well," and took the lift upstairs and waited. Eventually Sarkar came and said, "Let's sleep."

I replied, "I'm still waiting to meet Mrs. MacDonald."

"Is this any time to meet with someone?" he answered. "Come now, and meet with her tomorrow."

February 17, 1910: When we met in the morning, Mrs. MacDonald said, "My lady friend remembers you often. It's a shame that you agreed to meet her but never came down." I apologized and said, "It was my mistake. In my haste, I mistakenly understood that you would come upstairs, and so I waited for you there." We both agreed it was a pity and laid the matter to rest. We chatted about this and that until 9:30, when she took me out shopping. She selected a petticoat and a fine undershirt for me and then asked, "Do you have any more undershirts?"

"Yes, I have two others. This is quite enough."

Then, without informing me, she bought me a present, a beautiful silver box filled with sweets. At the time, I didn't know what she had bought it for.

We returned at 11:00 a.m. Sarkar was waiting for me. I went with Sarkar and we took a carriage to the Pitti Gallery and some other places.[14] It is raining a little today. We went into the Pitti Gallery and bought tickets. If you have an umbrella, you must check it with the doorman whether you want to or not. And then you must pay him for the service. We accordingly checked our umbrellas and entered. We had a good look around this picture gallery. The walls and ceilings were adorned with large and unique paintings of every possible kind. We looked at them for so long that my eyes became sore. There were many other visitors, most of whom used mirrors to view the ceilings. We spent an hour visiting each room but still did not see everything. There were so many rooms, and some of them we visited only briefly.

We returned to the hotel for lunch at one. Then I went upstairs and met with Mrs. MacDonald. This is when she gave me the box of sweets that she had bought for me. "Here is something to remind you of me. Please keep it." This box, made of European silver, was very beautiful. It had a picture painted onto the lid. I accepted it with thanks. I gave her two Hyderabadi rupees to remind her of me. She said, "I will have them made into a brooch and keep them with me always."

Sarkar went out again, but I was tired and so did not join him. Soon it was evening, dinnertime. Sarkar returned, got dressed, and went down to eat. I went to sit in the drawing room and wrote a few letters. Then we went to the room and slept.

February 18, 1910: We awoke in the morning as usual and had breakfast in the room. Then we quickly got ready and hired a carriage to visit more places. The carriage driver tricked us. "Come and visit the emperor's palace," he said, and we agreed. When we arrived, we learned that it was closed today. He did this mischief just to increase his fare.

Today we visited a museum but only saw three of the rooms. There were paintings here too. Sarkar told me the story of these paintings. "A Venetian boy once fell in love with a fairy who was more beautiful than even this." There's no need to repeat the story here. I looked closely at all the paintings. I didn't want to go anywhere else today. There were more galleries, but they were closed and are only opened by government order. Then we got into the carriage and went to the university. We looked at the buildings there, saw portraits of the instructors, and went back to the hotel.

We had lunch at 1:30, and then I went up to my room as usual. Sarkar went out to tour again but returned after a short while to get me. Then we went to another university. There were courses in session. Students were coming and going. Once again, the ceilings, the walls, and the vestibules were adorned with paintings of various people, all of them buck naked. We saw all the buildings from the inside and then took a short jaunt through the city. We returned to the hotel before sunset. We washed up and got ready for dinner, which we took at seven. After dinner I met with Mrs. MacDonald. She was very sad when I said, "I will be leaving tomorrow." We chatted for a while, and then I returned to my room and went to bed.

February 19, 1910: In the morning I packed my things because Sarkar told me that "tomorrow we will leave for Florence on the two p.m. train." But I did my packing today. Sarkar's things are still left to pack. Then I had breakfast and got dressed. We hired a carriage and went out to tour. We went to a church. Services [*namaz*] were being held. Some two hundred priests bearing lighted candles passed in a procession before paintings of Mary and Jesus. Then they split into two lines and stood still. The head priests recited something loudly, and then they all joined in and began to sing together beautifully. They sang for a long time, sometimes stopping and then starting up again. The services were still ongoing when we exited the church and returned to the carriage.

Back on the road, we saw a strange army bearing musical instruments and some type of special insignia. We learned that the priests here have some sort of leader. Ever year on the anniversary of his death, his disciples bring out this procession in his memory, as though they were commemorating their master's *urs* or celebrating his *barsi*.[15] In short, we saw quite a bit today too. In the evening we returned to the hotel. After dinner I met with Mrs. MacDonald. She is very sad that I will be leaving tomorrow.

Departure for Florence

February 20, 1910: I woke up early, packed for Sarkar, and got ready. Sarkar went out and returned at noon. We paid our bill and had lunch. We left by the 2:00 p.m. train. Mrs. MacDonald, the poor thing, cried when we left. I won't be able to forget how she cried. We enjoyed the views throughout the trip and arrived in Florence at eight in the evening. We went to the hotel. It is also nice. We got a room and went to rest and sleep. We had already eaten on the train, in the dining car.

February 21, 1910: We woke up early in the morning and had a bath. We got dressed and went down to breakfast. Then we hired a carriage to tour. We saw the picture gallery and the museum. We saw excavated graves and other objects that were found in them. Then we came back to the hotel for lunch. I stayed in the room, but Sarkar went out again and didn't return until evening. We had dinner and rested. I'm a bit tired of this journey now. I'd like to go home.

February 22, 1910: We woke up as usual. Sarkar said, "Let's go."

"I'm not going," I replied. "I'm not interested in this trip anymore."

"We only came here so that you could see it. If you won't see it, then we will have come here for nothing."

On his order, I got ready. We saw the famous places here. If you ask me the truth, this city is peerless. It is extremely beautiful, pure, and clean. However, although the churches look very grand from the outside, inside, they are shrouded in darkness. Today we returned to the hotel at 12:30. After lunch Sarkar said, "Let's go out again."

"Please, excuse me this time."

Sarkar went out but soon returned. "Come, I will show you the entire city in a motorcar. Then we will leave tomorrow."

I was delighted by the phrase "we will leave tomorrow" and thus happily agreed to the proposal. I went and sat in the motorcar. There were six other mems and sahibs with us. This is a large, open-top car. It is very nice. We toured in this car for three hours. I cannot praise this city enough. It is vibrant, clean, and beautiful. It is so nice that it is like heaven on earth. What beautiful gardens, squares, and fields, located high in the hills. We toured for three continuous hours, getting out to visit places once or twice. There's also a beautiful tunnel in the city. We returned to the hotel in the evening. We had dinner and went to bed.

February 23, 1910: We woke up early in the morning, had breakfast, got dressed, and packed our things. "We will leave on the same twelve o'clock train." We had an early lunch and left for the station. Here we ran into our old fellow travelers, our mem friends who had come from America and who had stayed with us in the Pera Palace, in Constantinople. We chatted for about ten minutes before boarding the same train. However, they are in second class, and we are in first. The two classes are close enough, though, that we can easily come and go from their carriage. We had a final meeting with them in the dining car. I proposed that "if I stay very long in Paris, we shall certainly meet there," because these mems are going to stop somewhere else first along the way.

We had to change trains at the same station as before. We made the transfer at nine. Our carriage in this train was completely empty besides us, and we slept well all along the way. At three in the morning Sarkar woke me up to show me a famous lake. There was so much snow on the ground that you could not even see the tracks. The mountains and the ground were all white. I took in this enchanting sight for a while and then went back to sleep.

In Paris

February 24: I woke up very early, washed up, got dressed, and watched the snow fall. Then Sarkar and I went to have breakfast in the dining car. When we returned, we found a French mem and sahib in our carriage. They had a conversation with Sarkar in French. We had lunch in the train too. At three thirty we arrived in Paris. Adopting a leisurely pace, we took a carriage to the Louvre Hotel.[16] Sarkar and his father have both been here before.[17] It is very nice. The lift is very large.

I washed up, got dressed, and had tea. Then we went to the [Grand] Magasins du Louvre, which is a large store directly across from the hotel. It was 5:30, and there was a rush of people. We bought a few hats for the children and placed orders to have a few things stitched for ourselves. We gave our names and then stayed in the store until seven. Then we went back to the hotel, had dinner, and went to bed.

On February 25, we had breakfast and went back to the same shop at just after nine. This shop is very large. I picked out some cloth and some lace and selected a design for a coat. Then we got a carriage and went to tour Paris. There was a huge storm here too, and many buildings had been washed away.[18] We had already seen the ruins of some of these buildings from the train.[19]

Paris is a very fine and expansive place. It has a large tower that you can ascend in a lift.[20] I did not want to go today on account of the rain. We walked around for a long time and then went to the museum. The museum was closed when we arrived, but we were told it would open at 2:00. Still, we had a good look at the building itself from the outside. After visiting so many museums, I can no longer muster any further interest; we have already been to hundreds of museums. From here we took a motorcar to the hotel and had lunch.[21] After lunch I did not go out again. After dinner I went to bed.

On February 26 I went back to the same shop again. They had made a rough fitting for the coat and set it aside. I tried it on and had the measurements amended. I asked them to correct the fitting now so that there would be no issues later. "We will be leaving by the eleven o'clock train tomorrow," I said. "Please have it sent to London by parcel." When I went into the shop today, they said, "When can we meet your husband?"

"You can meet him later this evening, or tomorrow he can stop by before we leave. You may speak to him then."

I went back to the hotel. Sarkar was waiting for me. We went to see the famous picture gallery and museum.[22] The picture gallery is nearby, so we went by foot. Today is a free day, so we went in without paying anything. The paintings inside were indeed impressive. The door frames were glorious. The walls and the ceiling are all adorned with paintings. The pictures are incredible, and the vastness of the building is beyond description. We began at ten and did not leave until three. There is also a small museum here, which we also visited. We returned to the hotel at three. We missed lunch today in the spectacle of it all. We made do with dinner and went to rest.

Leaving Paris for London

February 27: Sarkar and I are ready to depart. We only went out to the shop for ten minutes to ask them to quickly finish the sewing and send the clothes on to London. We gave them our address but then changed our mind and said, "Have it sent directly to your shop in London. We will collect it from there ourselves so that we can check the fitting." Then we went to the hotel to eat, pay our bill, and clear our account.[23]

We left on the 11:00 train and arrived at Calais Station four hours later.[24] From here it is a two-hour journey through a rough sea. So Sarkar and I, alongside about one hundred angrez, got off the train and boarded the ship. Immediately upon boarding, all the ladies went and lay down in their beds. Right away "maid stewards" came and placed a vessel for vomit beside each of their beds. I became worried because I have started to feel dizzy and nauseous at sea. I quickly went to lay down and closed my eyes. There were only women there; the men were elsewhere. I fell asleep and slept so profoundly that I only awoke when the ship arrived. I neither woke up nor felt dizzy nor vomited. When I did open my eyes, what did I see but all the poor women around me red-faced, their vessels filled with vomit. As soon as the ship stopped, Sarkar asked how I had fared. I replied, "Alhamdulillah, thankfully, I was quite comfortable." By God's command, Sarkar was comfortable too.

London is two hours from here. It was 6:00 p.m. We took a train. There were many angrez [men] and angrez women in our carriage. We were comfortable in our seats. We had our luggage stored in another carriage.

Arrival in London

We arrived in London at 8:30. We had our luggage removed from the train and then took a carriage to the Westminster Hotel.[25] No one met us at the station because we did not inform anyone of our arrival.[26] We mailed off three letters while still at the station. The first went to Usman Nawaz Jung Sajjad Beg, the second to Mirza Abbas Beg, and the third to Khwaja Abd ul-Majid.[27] This was so that all three would come to us first thing in the morning. Then we went to the hotel, had dinner, and rested.

February 28: In the morning I woke up, took a bath, got dressed, and had breakfast.[28] Sarkar went out somewhere. I sat to wait for Sajjad, Abbas, and Majid. At long last Majid appeared, at twelve o'clock. I was so happy to see him. I asked after Mahmud.[29] He replied, "He is doing well. I will ask him to come whenever you'd like. Just give the order. If I send a telegram today, he will be here tomorrow." We were still mid-conversation when lunch was served. We went to the table. I asked after Sajjad.[30] "He may well be in London these days, but Abbas is in Cambridge." I replied, "I wrote to him too."

Suddenly, a telegram from Abbas appeared. It read, "I am coming on the eleven o'clock train tomorrow." We were still eating when Sarkar came and joined us for lunch. Then we all had a long, lively conversation. Afterward, we went upstairs. Sarkar said, "Ask Mahmudullah to come on March 7." After this was settled, Sarkar went out again. Majid and I sat and chatted for a while longer over tea. Then it was time for dinner. I spent the day in conversation and didn't go out at all today.

We three were at dinner when a card came. It read, "Dr. Usman Nawaz Jung." The moment I saw the card, I leaped to my feet, for it had been eleven years since I had seen my brother Sajjad, and I wanted to meet him as soon as possible. I had already stood up from the table when Sarkar said, "Hold on, hold on, let him come first!" Sajjad came in, gave me a hug, and said, "Oh, *apa*! Oh, apa! Where have you been?!" He was just as happy to see me. Thank God, he has become a successful doctor. Sajjad had already eaten, so we finished our meal and went upstairs and sat together until midnight. Then Sajjad Beg and Majid left, and we went to sleep.

On the morning of March 1, we followed our usual routine. We got up, bathed, and dressed and then went down to breakfast. After eating I returned upstairs while Sarkar went out on business. Suddenly Sajjad and Majid appeared. "Come on, we want to give you a tour." I replied, "Wait, we'll go after lunch." We talked for a while. We went out after lunch. First, I wanted to go on a long walk. After that we took a taxi to Hyde Park and spent a long time looking around. This is a wonderful place for an outing. Next, we took a walk around Buckingham Palace (the monarch's palace) and looked around the city on our way back to the hotel. When we arrived, we found Abbas waiting for us. We all had tea and then dinner. Sarkar came as well. After eating we all went upstairs and chatted until 11:00 p.m. Abbas has only come for one night; he is staying in the room opposite ours. Sajjad and Majid both left, and we all went to rest.

A Tour of London

March 2: We woke up as usual and had breakfast. Abbas and I began to talk. Sarkar had a lunch appointment somewhere, so he went out. Majid and Sajjad came. Sajjad said, "Come on, apa, today I am going to make sure you get some fresh air. I am not going to let you sit here all shut up in a room. I'm taking you to see the museum and Westminster Abbey and God knows what else! The Zoological Garden. Basically, I am going to show you everything!" The four of us went out after lunch. First we saw Westminster Abbey, then Green Park, and the Zoological Garden.

Then they took me for a ride in the "Underground Rail." This is the train that runs beneath the ground. You descend underground via a lift. Inside it is very bright. Sometimes light comes in from outside, but for the most part the light and the climate are controlled by electricity. There is every type of animal in the Zoological Garden. The "rolves" danced and played of their own accord when they saw us. They dove in and out of the water.[31]

We had a good tour of everything there. Then we took the underground back to the hotel. After dinner Sajjad said, "All right, now I am going to take you to the theater." He insisted on this with *bhai* sahib [his brother-in-law, Nawab Sarbuland Jung] and then, after getting permission, took me away. Majid went home, and Abbas took a train to Cambridge. Sarkar went to rest.

Sajjad took me to the theater. I don't remember the name, exactly, but it was a wonderful show. I was thoroughly entertained. It ended at 11:00. Here the plays begin at 8:00 p.m. and finish at 11:00. We took a taxi home. Sajjad Beg dropped me off. He went home; I went to sleep.

March 3: I woke up, bathed, and had breakfast. Sajjad and Majid came, and after lunch, Sajjad declared, “Come, I am going to take you to see the shops and make sure that you get some fresh air.” Sarkar had to go out for lunch again. Before leaving, he told Sajjad, “I’ve arranged for tickets to the House of Commons, the House of Lords, and the House of India [the India Office]. Take her to see them.” Then he left, and we set off on our tour. Sajjad took me to a large shop. We had a good look around and then went on to another shop, where I bought a handbag for Father embossed with his name. Then we wandered here and there before returning to the hotel. They left at 11:00 p.m., and we went to sleep.

March 4: We woke up and had breakfast as usual. Sajjad and Majid arrived. Sarkar said, “Three invitations to tea have come for you today. I told them you would not have dinner. Those who have invited you are Lady Sir Charles Lyall, Lady Cromer, and the wife of some other Lyall as well.” Majid said, “At eight thirty, Ms. Beck will come to visit *bhabi jan* [Begum Sarbuland].” Sarkar left because he had a meeting somewhere. I went out with Sajjad. We had a short walk before returning. Ms. Beck came after dinner. We talked until eleven. Eventually Sajjad and Majid left. Ms. Beck did too. Then we slept.

An Invitation to Sir Charles’s Home

March 5: We woke up and ate as usual. Sarkar said, “Lady Sir Charles Lyall has invited you over at eleven. Please get ready.” Meanwhile, Sajjad and Majid showed up. Once I was ready they asked me, “When will you return?”

“I’ll be back in an hour or two.”

“All right, you go on, and we will go for a walk but come back before you do. Bhai sahib [Hamidullah] asked us to get him a trunk; we will try to find one.”

Sarkar and I took a taxi to the home of Sir Charles Lyall.[32] Lady Charles looked after us very attentively. She served excellent tea and plenty of cake. We had a long conversation. Both her daughters were there too. It was through her that I found the address of Ms. Lorrimer, who used to be second master of our zenana school.[33] I expressed my wish to meet her. Mrs. Sir Charles said, “I will write to her, and she will come to meet you.” I went back to the hotel two and a half hours later. Majid and Sajjad were waiting for me. Sajjad took me to the theater again after dinner. He brought me back at eleven, before leaving again. Sarkar was already in bed; I slept too.

March 6: I woke up and was having breakfast as usual when the waiter came and said, "There's a telephone call. They are asking for you." I had no idea who could be calling me.

"Perhaps it is someone looking for Sarkar?"

"No, Madame, they asked for you."

I went to see who it was. It was Abbas, calling from Cambridge. We had a long conversation. He said, "Apa, I'll come on the eighth for two days."

"Please do." The eighth is the day after tomorrow.

I returned upstairs. Sajjad and Majid had come. Sarkar has dinner invitations here every day, so he left. I went with Sajjad and Majid to the Louvre shop thinking that the clothing I had stitched must have arrived. When I arrived, I was told that it had indeed. I tried on the coat and found it fit nicely. I took the clothes and went back to the hotel. Today I was to go to Mrs. Cromer's house. I got ready at 4:00 p.m. I wore my new clothes. Sajjad and Majid said, "We'll wait for you to return."

Sarkar and I went to Lady Cromer's house. We had an excellent visit with lady sahiba. She gave us a wonderful tea. Lord Cromer came too.[34] He is an elderly man. We shook hands and talked. We chatted for an hour and a half before returning to the hotel.

We had dinner with Sajjad and Majid. Sajjad took me to the theater again. I really liked the show today. It was a magic show. When the Englishman performer wrapped a mem up in cloth and brought her close for me to examine, I pressed on the cloth and felt that there was something soft and cold inside. As soon as I touched it, it gasped in surprise. I laughed. A few moments later she reappeared elsewhere before disappearing into the cloth again. We came back at 11:00 o'clock. Sajjad Beg went home, and I went to sleep.

March 7: I woke up and had breakfast as usual. Majid and Sajjad came to the hotel. Just after Sarkar went out, a telegram arrived: "Mahmudullah comes today." It made me very happy. A lady sahiba whose name I have forgotten came to meet us today. When we received her card, Sajjad and I quickly went downstairs. Majid had gone to the station to collect Mahmudullah. I had a very nice visit with the lady. I offered her tea and ice [cream?]. We spoke for a very long time. She waited a long time for Mahmudullah to arrive. Eventually it got to be too late, and she had to leave.

He showed up a little while after she left. Sajjad was very happy to see him and gave him a big hug. Mahmud has completely forgotten Urdu.

He speaks English very well. I became concerned. I tell him, "Son, speak Urdu." Even I was compelled to speak to him in English. Mahmud's cheeks are turning pink here, as though he really has turned into a rose. He is very healthy, thank God. I was so happy to see him. Sajjad was overjoyed. He picked him up and seated him in his lap. Sometimes they wrestle with one another, and other times they chat together. After seeing me, Mahmud now says he wants to visit India, but his term is not yet up. I said, "I'll have you come back with Majid a month and a half from now."

By then it was dinnertime. Sarkar came back, and we all had dinner together. Sajjad took Mahmud and me to the theater again. I said, "Bhai, *bas*, enough, I've seen enough theater now. But fine, I'll go again today for Mahmud's sake." As usual, we returned to the hotel at eleven. Sajjad left, and we all went to rest.

March 8: We got up and had breakfast as usual. Sarkar went out. Majid and Sajjad came. Abbas came at eleven. We all went to have our picture taken. Majid left for class but planned to come back at four. I said, "All right, we can have the picture taken now or when Majid comes back." We agreed on the latter, and he left. Then the four of us went for a long walk and saw the underground rail tube several times, which has a large engineering museum, until it was time for the photograph. We arrived at the photographer's studio at three. Majid came, and we had our pictures taken (plate 7).

From there Abbas and Majid strolled back to the hotel, but Sajjad, Mahmud, and I took a taxi directly to the House of Commons. I went inside to the ladies' gallery and Mahdmudullah and Sajjad Beg to the outer, men's section. To get to the ladies' section, you take a lift up. There were many other ladies present when I arrived. Mrs. Holden, whose brother Mr. Baldwin was giving a speech, was seated near me.[35] She is a very affable lady. She looked after me very well and lent me a pair of binoculars: "You should use these," she said. She explained to me, "These people rule every region here in the House of Commons." She invited me to tea and gave me her address. I stayed until seven before taking my leave and going back downstairs. Sajjad and Mahmud were just exiting too. We went to the hotel, where we found Abbas and Majid. Sajjad said, "Apa is amazing. She makes friends wherever she goes." We had dinner. The Becks came to visit again. They are two sisters. We concluded the gathering at eleven.

March 9: I got up as usual. Abbas and Mahmud are sharing a room. They also got up. We all got ready and went down to "breakfast." Sajjad came [to the hotel] while we were eating. "Come, apa, I'm going to get you glasses made." Sarkar went off somewhere again.

An Invitation from the Prince of Wales

After Sarkar had already gone, I received a letter that read, "The Prince and Princess of Wales invite you to tea. They request an immediate reply." Sarkar wasn't even there. What was I supposed to do? I sent the attendant back, saying, "We will reply by telephone. You may go; we will reply as soon as Sarkar returns."

Then we went to the doctor. Sajjad had my name recorded as "Ms. Beg" to get a lower fee. He charges much higher fees to those named "Mrs." He will take a lower fee if he considers me Sajjad's sister. There are never fees for his wife or children; doctors don't take a fee from one another. After a careful examination, the doctor said, "You should definitely be wearing glasses," and wrote down the power. Then we went to a shop for eyeglasses and placed an order to have glasses made. "A pair of glasses will cost two pounds and two shillings," they said. We finalized the order and then went to the hotel. We all had tea together. We were still having tea when Nawab Sarbuland Jung sahib bahadur returned. I gave him the prince's letter. He immediately sent a reply: "We will certainly visit tomorrow." We all had dinner, and as usual, Sajjad and Majid left at eleven, and we went to sleep.

On the morning of March 10, we awoke, got ready, and had breakfast. Abbas must go to Cambridge today. Ever since he and Mahmud came, we have been staying in a suite with two bedrooms, a bathroom, and a drawing room, everything. Sajjad came. Majid will come at three, after class. Sarkar left before lunch but told Sajjad, "Miyan, I have lunch out today. Mr. Higg will come to collect the begum. She should be ready to go by three on the dot. You should make sure that she and Mahmud leave with him. And please tell him that I will go there directly from where I am."

After lunch we sat and talked. At two, Sajjad said, "You should go and get ready now." I washed up and got dressed. Mahmud also got ready. A waiter brought a card to notify me that Mr. Higg had come. The three of us went down; Abbas stayed [in the hotel room], saying, "I'll leave at nine."

Mahmud and I went and set off in a taxi with Mr. Higg. Sajjad and he had a long conversation. He is a good man; I met his wife twice in the Hyderabad Residency. He speaks Urdu fluently. He expressed a wish to "[be appointed as] resident and move to Hyderabad." He really liked Sajjad. He told me, "Your brother is a good man." I agreed. We chatted in this manner until we arrived at the palace. We got down from the taxi. There is quite a guard system here. Each of them doffed their hats for us, and we went inside.

We passed through a wide courtyard before entering a room. Sarkar was already there. He spoke to Mr. Higg. There was another gentleman there too. All the servants were standing in a line; they too removed their hats. We went into the hall and took our seats by the fireplace. The Prince and Princess [of Wales] entered five minutes after I arrived.[36] I stood up. The prince shook my hand first and then Sarkar's. Then the princess shook Sarkar's hand and then mine. The princess sat down and had me sit near her. The Prince of Wales remained standing while speaking to Sarkar. The princess chatted with Mahmudullah. "This boy speaks English well," she said.

"He is going to school here [in England]," I replied.

"I still remember the poem that you wrote about me. Could you give me a translation of it?"

"Of course. As soon as I have returned to India I will have a translation made and send it to you."

She then began to tell me about her children. "All my children are in school. I have six, five boys and one girl. I really liked Hyderabad. If you meet the nizam, please give him my greetings. And if for whatever reason you are unable to do that, then please give my greetings to his wife instead."

"Of course. I will be sure to give them your greetings."

Then the prince began to speak to me, and the princess went to speak to Sarkar. He said, "You must be sure to have a translation of your poem made and send it to me."

"Of course."

"How do you like London?"

"I like it very much," I replied.

"India is very nice."

"It also belongs to you," I answered.

We all spoke together for an hour before they left for another room and we went outside. All the attendants doffed their hats as we went.

When we reached the gate, a soldier immediately called for a taxi. Mr. Higg accompanied us. The method for calling a taxi here is that you just whistle, and a taxi appears. We said goodbye to Mr. Higg and left. It was dinnertime when we arrived at the hotel. We quickly gave Abbas his dinner. We also ate and then went upstairs. We sent Abbas off with Majid, who took him to the train station before returning. We all talked until eleven, when Sarkar said, "Please, go to sleep now, both of you. You will fall ill if you continue to stay up so late every night." Sajjad and Majid left, and we went to sleep.

March 11: We woke up, got ready as usual, and had breakfast. Sarkar had a lunch to attend today too, so he left. Sajjad, Majid, and I went out for a while to take the air. Sajjad Beg got us an excellent trunk and had another trunk of leather made for our belongings. Sajjad Beg and Abd ul-Majid packed up all our things. The glasses I ordered have come too. Sajjad had two pairs made in case the first should break. We spent the entire day in these tasks. As usual, we had dinner, and they went home.

March 12: This morning Sarkar left at 4:00 a.m. to meet with his friend Sir Fitterzel in Plymouth. Before leaving, he said, "I'll be back at three in the morning. We leave London at nine [tomorrow]. Please be ready." Sajjad and Majid have been here since the crack of dawn. I went to Mrs. Holden's and left my card. Then I went to Mrs. Ameer Ali's because it is raining today.[37] I also packed up Mahmudullah's things. Sajjad and Majid stayed up until 1:00 a.m, and we had a very enjoyable conversation until then. I pressed Sajjad to "come back to India as soon as possible. Don't think about traveling abroad any longer." He, however, passionately wants to visit Constantinople. His brother-in-law [Hamidullah, presumably] has given him several letters of introduction. He's already arranged his passport. Still, I tried to explain things to him. I hope he understands. After all, he is very intelligent, capable, and well-mannered, masha Allah. Inshallah, Majid will pass his exams and bring Mahmudullah back with him in a month and a half or so, God willing. Masha Allah, Abbas has advanced so much, so quickly; it made me very happy to see. The two left at 1:00 a.m. "We will come again in the morning." Sarkar returned at 4:00 a.m.

Notes

1. They presumably docked at Piraeus.
2. The Grand Hotel d'Angleterre was in Syntagma Square, in the heart of Athens.
3. This certainly refers to the Parthenon.
4. The dargah of Baba Sharaf ud-Din is located just outside the city of Hyderabad. She had visited it in the first chapter of this travel account.
5. This almost certainly refers to a book in either the Baedeker Guide or Murray's Handbook for Travellers series of travel guides. It may refer to the fourth edition of the Baedeker travel guide to Greece, which was published in 1909.
6. The diary gives "Karinta." This and other errors are likely introduced by the lithographer rather than by Begum Sarbuland herself.
7. The Corinth Canal, completed in 1881, is about twenty meters wide. Here she uses the English word "lake."

8. The diary reads "Pairas" here and elsewhere, but the passage clearly refers to Patras.

9. The Thomas Cook and Co. representative.

10. This hotel was in the center of Rome, just a few blocks from Rome's central terminal. She calls it "The Kurital," but this may be the result of a scribal error.

11. Unclear.

12. She refers to the legendary Romulus and Remus, though records both as being given the name "Roma."

13. A museum guide, perhaps.

14. More commonly known as the Palazzo Pitti.

15. The term *urs*, literally "wedding," is a Sufi term that refers to the death day of a saint or holy person. It is so called because that date marks their return to and union with God. *Barsi* is another word for an annual commemoration of a death.

16. Though she gives the name in English, this is in fact the Hôtel du Louvre, located just across the street from the Louvre Museum.

17. The phrase used here for father is *abba jan*. As Begum Sarbuland habitually refers to her own father as *bhai abba jan*, and because there is no record of Agha Mirza ever visiting Europe, this presumably refers to Hamidullah's father, Samiullah. Father and son had traveled to Europe together in 1880. For a description of the trip from his father's account, see S. Khan, *Musafiran*.

18. The word "too" seems to be a reference to the Great Musi Flood of 1908 in Hyderabad. She seems to connect her own experience with the floods to the more recent deluge in Paris.

19. The 1910 Great Flood of Paris. A month before her visit, the Seine had reached to nearly thirty feet above its normal height.

20. The Eiffel Tower—which clearly did not impress her—had been intended to be demolished the previous year, in 1909.

21. The museum they visited was presumably not the Louvre; otherwise, they would not have needed to take a motorcar back to the hotel.

22. This was presumably the Louvre Museum.

23. It is noteworthy that in this European portion of the trip, Begum Sarbuland has stopped recording her daily prayers.

24. This line appears somewhat humorously in Urdu, albeit unintentionally. The phrase "at Calais Station" is rendered such that it could be read as "kele ke station par," which, in Urdu could mean "at the banana station."

25. This was most likely the Westminster Palace Hotel, a luxury hotel located across from Westminster Abbey. Mohandas Gandhi had stayed in the same hotel a few months before their arrival.

26. The couple had a large circle of friends and family in London who might have met them.

27. The first two men were her brothers. The third was her husband's maternal cousin. Abdul Majeed Khwaja (1885–1962) was the son of Khwaja Muhammad Yusuf, an early and ardent supporter of the Aligarh movement and a close friend and the brother-in-law of Hamidullah's father, Samiullah Khan. He studied at Cambridge University between 1906 and 1910. He reportedly studied there to become a barrister to earn the hand of Begum Sarbuland's daughter Khurshid Begum. He was thus in some sense auditioning for the role of son-in-law as he hosted Begum Sarbuland in London. After returning to India, he practiced law in Aligarh and, later, Patna. He gave up law in 1919 to support the anticolonial movement and to help establish Jamia Millia University in Delhi. He then returned to law, practicing in Allahabad between 1926 and 1944. He retired to Aligarh.

28. This sentence is indicative of Begum Sarbuland's occasional but surprising use of English in the diary. Rather than use standard Urdu words, in this instance she writes "dress kar ke breakfast khaya."

29. The author's middle son, aged eight, who was studying at a boarding school in Plymouth.

30. The author's brother.

31. The cute animal being referenced here is unclear. This may also be the result of a scribal error.

32. Sir Charles Lyall (1845–1920) was an Arabic scholar and a civil servant in British India. He probably did not work directly with Hamidullah. Instead, the pair may have known each other through their shared interest in the Arabic language. His wife was Florence Fraser.

33. That is, Mahbubiya Girls' School, where her daughter Khurshid Begum was studying.

34. Lord Cromer, or Evelyn Baring (1841–1917), was a powerful colonial administrator who made his career in Egypt. He was cousin to the Earl of Northbrook, whom Hamidullah had served as a private secretary in Egypt in 1884.

35. Probably Stanley Baldwin (1867–1947), who would thrice serve as prime minister.

36. The Prince of Wales, George Frederick Ernest Albert, had toured India in 1905–6, during which time he seems to have met the Sarbulands. He became king in May 1910 as King George V. He was coronated in London in 1911. He then traveled to India to attend the Delhi Durbar of 1911, where he was presented to a mass gathering of Indian rulers as Emperor of India. He was joined in his travels by the Princess of Wales, Mary of Teck, or, after 1910, Queen Mary. The Sarbulands were in attendance.

37. Isabelle Ida Konstam was the wife of Syed Ameer Ali, a prominent Indian jurist, writer, and political leader. Around the time of this visit, he was heavily involved in organizing Muslim political and social life in London.

CHAPTER 10

HYDERABAD

حیدرآباد

Returning to India

We were completely ready [to leave] by the morning of March 12.[1] Mahmudullah had already woken up and had breakfast, but neither Majid nor Sajjad showed up, so we left and went to the station. No one was there either. We had left his [Mahmud's] luggage at the hotel since he needs to go to Plymouth. I had thought to myself, "We will surely find Sajjad and Majid at the station. I will send Mahmudullah with them." When no one appeared and it was nearly time for the train to leave, we wrote the hotel's address on a card for Mahmud and left him at the station [to make his way back alone]. I couldn't stop thinking about how we had failed to meet them. We even sent a telegram though we were already on the way: "Please take care of Mahmud." We sent another to the hotel: "Please look after the boy carefully." These worries gnawed at me for an entire day and night.

Leaving London

Today is March 13. I have been worrying about Mahmudullah, Sajjad, and Majid since last night. Why hadn't they come? In any case, we boarded the *Gharbia* in the port and set off for India, but I continued to worry.[2] I thought about it the entire night.

March 14: The seas are rough. I feel nauseous. The mems and I vomited several times. I struggled through and was better by evening. The ship's doctor gave me some medicine, and the seas also calmed a bit. By God's grace, we made it through the night.

March 15: Today was a very pleasant day too.[3] I met a Parsi lady. This time all the English people aboard the ship have lived in India for years. I didn't like their manners at all.[4] Even if I have generally found European men and women to be well-mannered, it is still true that "God does not make all five fingers alike."[5] Every nation has both good and bad people.

March 16: We reached Port Said at twelve today. Many poor Arabs brought many things [aboard] to sell. I spoke to them at length in Arabic. All the ladies were shocked. I told them, "We are Muslim. Our pure Qur'an is in Arabic. We know Arabic very well."

The ship will remain here from noon until 4:00 in the morning. It will depart at 4:00 a.m. The ship is waiting for the European mail to be brought from Brindisi. The buying and selling on board the ship continued until 9:00 p.m., but we went to bed after dinner.

March 17: I thought that the European mail must have arrived last night and that there would surely be a letter from Majid. Majid's letter soon turned up. I was very happy to know that he and Mahmudullah were all right. He wrote that he and Sajjad had gone home together but then overslept, and that is why they were late. We left the hotel at nine, and they showed up there after. Mahmudullah had already managed to return to the hotel before they arrived, masha Allah. Majid sent him off to school and arranged everything for him there in advance by telegram: "Mahmudullah is coming." He was going to Plymouth, which is only a few stations away. I am so glad to know that all is well. I am a bit more at peace now.

Returning Home

Translator's note: The following pages appear to be a summary of the journey from London written at a later time and were not included in Begum Sarbuland's original diary. She had clearly tired of making entries by the end of the long trip and here this section brings the travel account to a quick close.

These tears of mine turn into pearls
When your sight appears before my tear-moistened eyes

I woke up, prayed, had a bath, got ready, and had breakfast [at the hotel in London]. The luggage was packed and ready. I had it removed from the room and sent on to the station. We were planning to leave for the station after Khwaja Abd ul-Majid and my brother Usman Fateh Nawaz Jung bahadur came so that I could entrust to Khwaja Abd ul-Majid the light of my eyes, Muhammad Mahmudullah Khan, who was with me and who was only eight years old.[6] However, after quite a long time passed and neither of them showed up, we took Mahmudullah with us to the station. We arrived there at nine and waited for them some more. When they still didn't come and it was time for the train to depart, we had no other choice but to write down the name and address of the hotel and give [the paper] to the light of my eyes, Muhammad Mahmud Khan. "You take a Victoria [carriage] and go to the hotel. *Mamu* [uncle] and bhai will meet you there," we said. We did this, but my heart was racing the whole time. I felt ill at ease leaving a child at the station like that. I felt so bad I cried, but I had no other choice.

The train left London around ten. The trip passed in worry and vexation. We arrived at ten. I met some American ladies and spoke with them for a while. We arrived at Dover around one and had lunch. Then we boarded a ship and crossed the same channel again. We arrived in Paris around three and had the baggage transferred from the ship to the train.[7] I wandered about for a bit. We had dinner in a restaurant at eight in the evening. Sarkar said, "I'll go change some pounds. You go to the train and sit in the cabin." It was almost time to leave. I went to the station by myself. The train was very long. I walked along the train from one end to the other but couldn't find my carriage. I started to worry. God! It's nearly time for the train to leave, and I can't find my carriage nor my luggage. What if Sarkar assumes his begum is already in her seat and he arrives at the last minute and quickly boards another carriage, leaving me behind? I am nearsighted, so I will not be able to see him when he comes [to board the train].[8] It is a full moon tonight, and all the Europeans will be wearing black suits, and Sarkar is wearing the very same kind of black suit. How will I recognize him? Tears welled in my eyes. To make matters worse, in Paris they speak French, which I do not know. Even if I wanted to ask someone for help, they wouldn't be able to understand English. I stood there thinking about all of this when suddenly I saw Sarkar in the distance calling out to me as he approached. "Run! The train's already moving!" I set off at a run the moment I heard him. Thankfully, we found the correct carriage. We took our seats.

"What happened to you? Why were you just standing there in the station?" he asked.

"I couldn't find the right carriage. When we came to the train earlier, there were just one or two carriages, and our luggage was inside. But when I came back, the train was much longer, and I could not locate my carriage. I was nervous and got scared that you would be in a rush and board the train via another carriage, assuming I was already on the train. So I just stood there crying. Then I thought that if I were to board the train now and you end up being late, then I'd have set off by myself. So that's why I just stood there on the platform."

Anyhow, we arrived in Brindisi at six in the morning. We got a ship from there from the very same P&O Company that was about to leave for India. I was returning home, but all I could think about was Mahmudullah. I've made friends with the wife of a doctor who lives in Mussoorie.

We reached Suez on the fourth day.[9] I had been awaiting our arrival here and hoping there would be mail waiting from London so that I could find out what happened with Mahmudullah. But there was no mail. Alhamdulillah, when we reached Port Said there was mail at last. Khwaja [Abd ul-Majid] and my brother wrote that they were sad to have arrived late. They reached the station in time to watch the train pull away before their eyes. "We did not meet you, but we took Mahmudullah with us and sent him on to Plymouth School the very same day. Don't worry." I thanked God.

At Port Said, Arabs brought many goods for sale [onto the ship]. I bought some socks and two *kamdani* shawls. Then the boat set off again and arrived in Aden after four days. Four days later it reached Bombay. We disembarked here. Professor Muin ud-Din met us at the station and took us to his house. He kept us fed all day long. I went to meet a relative of Lady Hydari's.[10] I bought a few more things from Bombay, and then we both went to the station in the evening to take the train to Hyderabad Deccan. We arrived the following afternoon at four.

In Hyderabad

Sarkar instructed me to leave the train at Begumpet.[11] "I already sent a wire. There will be transport waiting. There will be a huge crowd of government officials and my friends at Hyderabad Station. It will be a hassle for you." I got down the moment we arrived at Begumpet Station. Lady Doctor, Lady Hydari, and my mother had come to the station to receive me. My middle sister-in-law, Begum Doctor Usman Nawaz Jung,[12] and Sheila Walker were there too. I was wearing a Turkish burqa with a black niqab. When I flipped the niqab back, they saw that my face had been changed by the blessings of my pilgrimage to the Holy Places and by the European climate. It now had a rosy tinge. Sheila Walker immediately said, "Are you wearing blush?"

"I'm not. Look closer. This is the light [*nur*] given by God, which was granted to me by my visit to the Prophet's court."

I easily arrived home with my mother and sister-in-law. My mother and all my siblings lifted the evil eye from us both. By the grace of God and the intercession of the Lord of Medina [the Prophet Muhammad], my family has looked after me so well and been so gracious to me. Once I was home and reunited with my children, my heart felt light and happy. I was at peace once I had returned to kiss the feet of my dearest mother. I offer every sacrifice to the Lord of Medina, who allowed me to make this journey through the East and the West. I sacrifice myself to the Lord of the Two Worlds. By presenting myself at the threshold of his grave, I had the honor to travel.

Notes

1. This date is repeated in the diary twice. This final chapter contains several inconsistencies.
2. This seems to suggest she sailed directly from England to India. Later entries in the diary, however, more convincingly state that she traveled to Brindisi by rail before boarding a ship bound for Bombay.
3. The diary entries here are somewhat confusing. Her use of the word "too" seems to be an error.
4. Indian travelers often remarked on the difference in attitude and behavior they encountered between long-term British residents in colonial India and that of the British in Britain, where social interactions with Indians were not too strictly policed and racial hierarchies were somewhat less rigid.
5. She uses a Persian phrase here: *Khuda panj angusht yaksañ na kard.*
6. Here she uses different names for Sajjad and Majid.
7. She presumably means Calais rather than Paris.
8. Clearly she had not started wearing the glasses that she had bought the week before!
9. Suez and Port Said seem to have been flipped in this paragraph. Her ship would have arrived at Port Said before passing through the canal to reach Suez.
10. This line is unclear.
11. The train's penultimate stop.
12. Sajjad's wife.

PART 2

Begum Sarbuland

A Life Untold

Akhtar un-Nisa Begum Sarbuland Jung was born in 1876, in the southern Indian princely state of Hyderabad. She died at age eighty-two in Delhi, the capital of independent India. A descendant of the Mughal line and the daughter of a powerful statesman, she spent her childhood in comfort at the heart of one of the world's richest kingdoms. At the age of eighteen, she married an up-and-coming young lawyer from Allahabad who represented the vanguard of the English-speaking Indian Muslim elite. Hamidullah Khan (who was later given the title Sarbuland Jung) first studied at a newly opened school for Muslim boys in Aligarh in India and then, for most of his teenage years, at Cambridge University. His union with Akhtar un-Nisa epitomized Muslim modernity in colonial India. If she was the embodiment of Mughal aristocracy and princely rule, her husband was an exemplar of a transformed and increasingly Anglicized Muslim intellectual social class that emerged in response to colonial rule.

After their marriage, Hamidullah Khan moved to Hyderabad to begin a family and work as a high court judge, a position he was assured of by dint of his elite education, linguistic versatility, and his father-in-law's influence. Begum Sarbuland spent the first years of her marriage in a state of almost constant pregnancy, giving birth to at least thirteen children over a period of about fifteen years. Despite this, she was never confined to her home. In Hyderabad she often went out with her relatives and friends to visit a Sufi shrine or to picnic in the countryside. Begum Sarbuland expanded her education by attending high school with her eldest daughter, learning English, and supplementing her childhood schooling by later experiencing a formal Western-style education. Her adventures outside the home were facilitated by a team of ayahs, wet nurses, and maids who helped look after the home and assisted in nursing and raising her children. She also benefited from the presence of a phalanx of siblings and relatives to provide support and sociality. In this way, Begum Sarbuland found sufficient time to pursue her interests in Islamic theology, Sufism, literature, and languages, among other intellectual pursuits.

Begum Sarbuland was also an avid traveler. Her first long-distance voyage may have been in 1894 when she went to Delhi to be married in her father's childhood home. Her first trip outside India was in 1909 when she visited the Middle East and Europe. The pace of her travels increased later in life. After her husband's retirement in 1912, she shuttled constantly between Hyderabad, Delhi, Bombay, Allahabad, and Shimla, to visit friends and family. For decades she was a regular participant at reformist women's conferences across South Asia. She sailed abroad several times to stay with her children in Europe, visiting Medina on each occasion. As these excursions suggest, Begum Sarbuland was not a timid woman. She was, as her granddaughter recalled, a colossus, "a queen in the family," her force of will irrefutable.[1]

This chapter offers a brief study of her life and further context for understanding her private travel account. It also includes a short biography of her husband, Hamidullah Khan Nawab Sarbuland Jung, the diary's second protagonist. Given his fame and extensive publications, Hamidullah's biography is not difficult to reconstruct. Begum Sarbuland's life, though, is almost completely undocumented. Unlike some of her contemporaries in Hyderabad, Begum Sarbuland declined to become a public figure.[2] Her male companions and contemporaries also probably avoided speaking about the details of her life in keeping with conceptions of respectability among elite Indian Muslims of the time. These factors make writing a history of Begum Sarbuland a daunting challenge, since her life can effectively only be recuperated by collecting scattered references to her in historical sources or by reading between the lines of biographical and autobiographical materials about the men in her life (those of her father and husband, primarily). In the following pages I rely on these historical sources, on the information given in her own travel diary, and, finally, on interviews with her surviving descendants, some of whom lived with her for many years. I also draw on private family archives that include photographs and other unpublished writing by Begum Sarbuland. Together, these sources shed light on her life and her diary and offer an instructive look at the life of a wealthy Muslim woman in late colonial India.

Ancestral Roots in Old Delhi

Begum Sarbuland—or Akhtar un-Nisa, the name her parents gave her—was born in Hyderabad but had familial ties to Delhi, the nominal capital of the Mughal Empire, and to the royal court at the heart of the city. Her great-great-great-grandfather was the Mughal emperor Shah Alam II (1728–1806). Though there were hundreds who could boast such a lineage—the emperor had over fifty documented children—this ancestry gave Begum Sarbuland's family an enduring connection to Mughal life in Delhi.[3] Despite falling on very hard times by the eighteenth century, the city and its royal court still commanded immense prestige across India. One's proximity to Mughal Delhi was considered a metric for assessing one's refinement and gentility. So powerful was this association that, as far as prestige went, it hardly mattered that by Shah Alam II's day the Mughal Empire had become impotent, its influence famously going not much further than the suburb of Palam. Begum Sarbuland's forefathers leveraged their association with the royal court to find gainful employment outside Delhi, a city that was becoming increasingly difficult to live in.

As Delhi declined, the British Empire in India was expanding in the east. Britain's growth came in distinct stages. In the eighteenth century and the first half of the nineteenth, the East India Company often sought to grow its influence largely by adapting to India's social and governmental institutions and hollowing them out from the inside. Rather than conquering indigenous states by force of arms, they often allowed pliant rulers to retain their thrones on the condition that they were willing to be "guided" by the British advisers, or "residents," posted to their courts. Even where they ruled directly, the British preserved the administrative languages and procedures of precolonial India, conducting much of their official business not in English but in Persian, Urdu, or regional languages like Bengali or Tamil.[4] Thus, despite the huge shifts in power taking place, the transition to colonial rule was accompanied by some sense of continuity. Men educated in Persian, Arabic, and Urdu were still able to find employers who valued and needed their skills. Nevertheless, British norms and Western education were slowly seeping in at the seams.

Then came 1857, when a revolt among Indian troops serving in the British military spread to the general populace and lead to a violent uprising against colonial rule across India. Delhi saw some of the most intense fighting. British troops besieged the city for months before ultimately putting down the revolt. Their victory was accompanied by an uncontrolled rampage as troops flooded into the city, looting and shooting on sight. The restoration of British control did not stop the killings but instead led to sham trials and summary executions. Thousands fled. Unable to reenter Delhi, many camped outside its walls. The victorious forces initially banned all Indians from returning to the city; later the ban was amended to prohibit only Muslims, whom the British distrusted and accused of fomenting the revolt. The Mughal Empire had come to an end. Among those camped outside the city, forbidden to return to their ancestral home, were Begum Sarbuland's nine-year-old father and his family.

Exiled from their hometown, unable to access the wealth and the prestige associated with their social status in Delhi, the family adapted: they learned English, familiarized themselves with British culture, and took posts in the colonial administration. Simultaneously, they continued to cultivate and draw on their long-standing connections to the prestigious cultural world of old Delhi. This meant emphasizing their affiliation with Delhi's royal codes of etiquette and its authoritative dialect of Urdu. This was a valuable claim in British India, where Urdu was by then the most common language of administration. This cultural capital and their willingness to adapt meant that Begum Sarbuland's family rose to even greater prominence after 1857 than before.

Agha Mirza Beg in Hyderabad

The pivotal figure in this transition was Begum Sarbuland's father, Nawab Agha Mirza Beg Khan (1848–1933). Born in Delhi in 1848, he was just old enough to remember life in Mughal Delhi but young enough to have his education shaped around the new realities of British India. His life was spent straddling and negotiating these two cultural worlds—a fact that would surely facilitate Begum Sarbuland's own comfortable familiarity with many aspects of princely and British life. Agha Mirza passed his early years in his natal home in the Farash Khana neighborhood of Shahjahanabad (today better known as Old Delhi). His house was just a few minutes' walk from that of the famed poet Mirza Asadullah Ghalib, who was, in fact, a relative by blood and marriage.[5] Agha Mirza's mother, Munavvar Zamani Begum, was more educated than most women of her time. She received a thorough religious education at home, having read the Qur'an in Arabic and studied it through Shah Abd ul-Qadir's (d. 1815) famous Urdu translation and commentary.[6] She herself taught Agha Mirza to understand the Qur'an through this translation. Agha Mirza's father, Javvad Beg (also known as Mughal Beg), had been a scholar of Persian, Arabic, and mathematics. If not for the 1857 rebellion, Agha Mirza might have followed a similar path.

After fleeing Delhi, Agha Mirza and his family initially hid out in the ruins around the city (in the region of today's South Delhi) until a relative located them and took them to the nearby kingdom of Alwar, where Agha Mirza's maternal uncle was prime minister and regent. With the entire country still in disarray, this was a harrowing journey. On the way, they were joined by some cousins, one of whom gave birth while they were still on the road. The newborn, Sikandar Zamani Begum (b. 1857), would one day become Agha Mirza's wife and Begum Sarbuland's mother.[7] Unfortunately, the group found little respite in Alwar, where a court dispute between rival Hindu and Muslim factions resulted in them being chased from the city. Exiled once more, the family attempted to return to Delhi but were still refused entry. So they turned east, traveling to Sitapur, near Lucknow. There they found safety and security with Agha Mirza Beg's paternal uncle Mirza Abbas Beg.

This was a turning point in Agha Mirza's life. In his early youth Agha Mirza had studied Urdu, the Qur'an, some Persian, and possibly some English. In Lucknow, though, his uncle introduced him to a British curriculum. Mirza Abbas Beg had come under the influence of the famed pedagogue Master Ramchandra, who famously supported Western education.[8] Guided by Master Ramchandra, Mirza Abbas learned English and joined the British Army, helping to conquer Punjab in 1849.[9] Then, in 1857, he defended the British at Lucknow. This display of loyalty earned him immense trust and favor in the city.[10] Through his British contacts, he had Agha Mirza enrolled in Lucknow's exclusive Canning College, an institution founded to teach a Western education to the Indian elite.[11] All the instructors at Canning College were British apart from the Persian and Arabic teachers. Soon, Agha Mirza was completely fluent in English. He spent his leisure time studying philosophers like Locke and Hume and reading English novelists, Sir Walter Scott particularly. (For an example of his English writing, see fig. 11.1.) He was also well read in Persian, Arabic, and Urdu.

After graduation, Agha Mirza levied his connections to find a post in Hyderabad. Hyderabad had sided with the British in 1857, a decision that spared it the brutal retribution meted to cities like Delhi and Lucknow. Thus, even as Delhi underwent a painful transformation to direct colonial rule by a vengeful colonial state, Hyderabad hardly suffered at all in the

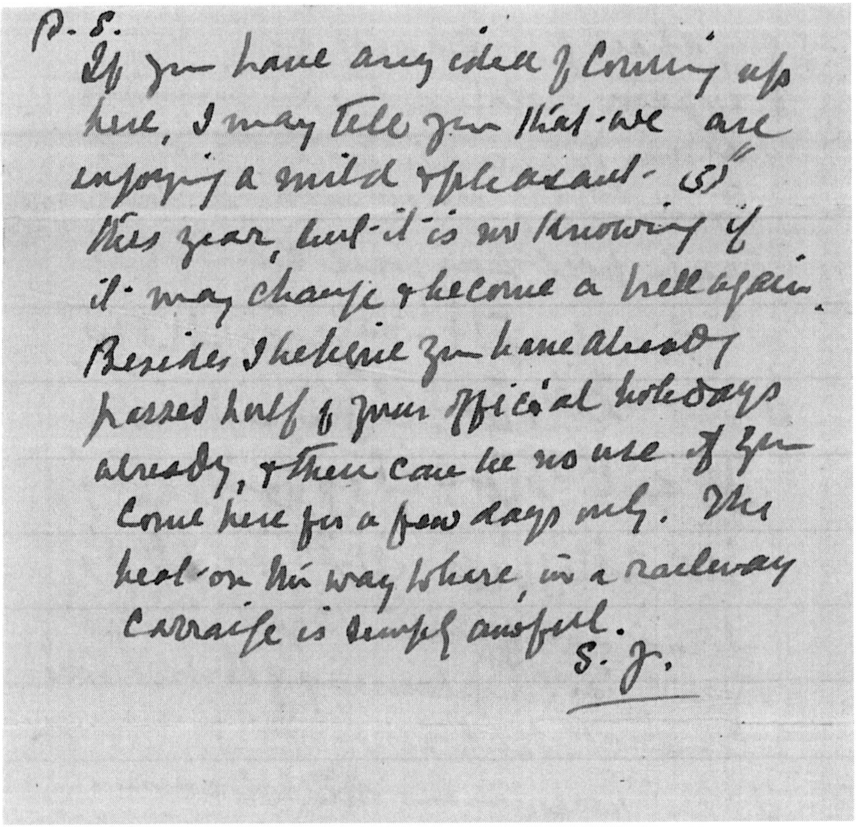

P.S.
If you have any idea of coming up here, I may tell you that we are enjoying a mild & pleasant [illegible] this year, but it is no knowing if it may change & become a hell again. Besides I believe you have already passed half of your official holidays already, & there can be no use of you come here for a few days only. The heat on the way here, in a railway carriage is simply awful.
S. J.

Fig. 11.1 The postscript of a letter written by Agha Mirza Beg to his son Haider Jivan Beg from Lucknow in 1919. *Source: Shakira Sultana.*

years after the uprising. In fact, it prospered. The wealthy, stable state easily attracted huge numbers of highly accomplished men from the north who came there fleeing violence and discrimination in their homelands. Among the migrants was Agha Mirza, for whom the pull of Hyderabad was surely irresistible. A rich and peaceful kingdom, Hyderabad maintained Mughal traditions, including the use of Persian as an administrative language. The Mughal influence in Hyderabad made Agha Mirza appealing to the state too. He was a descendent of the Mughal line; he spoke Delhi Urdu and was skilled in Persian. He had been trained in the norms of the Mughal courts from his infancy. At the same time, he was fluent in English and adept at maneuvering among British officials in India. He was a perfect match for a semi-independent state seeking to retain its traditions while adjusting to the new realities of life in high colonial India.

Agha Mirza arrived in Hyderabad in early 1873.[12] There he was appointed as tutor for the children of the prime minister and regent, Salar Jung I (1829–1883).[13] Soon thereafter he became the private tutor of the ruler of Hyderabad himself, Nizam Mir Mahbub Ali Khan, who was a young child at the time. Agha Mirza's knowledge of courtly life and his ability to move seamlessly among English, Persian, Arabic, and Urdu made him an ideal appointee for managing an education that straddled British and Indian curriculums.[14] As a result of this constant and intimate contact, teacher and student formed a close bond that lasted for almost two decades. Once his schooling had concluded, the nizam retained Agha Mirza as his personal secretary, adviser, and confidant.[15] They were so close that for some two decades Begum Sarbuland's father "exercised an almost unbounded autocracy" in Hyderabad until his retirement in 1897 (plates 2 and 3).[16]

Many of Agha Mirza's family members followed him to Hyderabad, including his widowed mother, Munavvar Zamani. In 1874, he traveled back to Delhi to marry his cousin, Sikandar Zamani Begum, whose birth he had witnessed in 1857 (plate 4).[17] Their marriage reaffirmed and expanded his ties to princely India: Sikandar Zamani's mother belonged to the royal family of Loharu, while her father had been prime minister of Alwar (which meant that Begum Sarbuland herself had immediate ties to these courts). Together the couple produced ten children.[18] All were born in Hyderabad but retained a social identity as northerners rather than as native Hyderabadis.[19] Begum Sarbuland was probably their first or second child. They named her Akhtar un-Nisa, meaning "a star among women." As a young girl she played with the children of Hyderabad's elite and surely met many of the influential figures from Hyderabad and northern India who visited their home.

In this way, for Begum Sarbuland and her family, 1857 was ultimately a moment of ascent. Her immediate family now occupied positions of more power and wealth in Hyderabad than they ever had in Mughal Delhi or than they might have achieved in British India, which denied Indians roles in the upper administration. Agha Mirza's success became generational, as his sons and sons-in-law took important positions in the state and were themselves awarded honors and titles. Yet this did not mean that success was guaranteed in perpetuity. In the end, Agha Mirza was forced out of the city, where political intrigue was rampant. By the 1880s, there was significant conflict between two major factions, the *mulkis* (those who had ancestral links to the city) and the *ghair-mulkis* (those, like Agha Mirza, whose families had migrated there).[20] These disputes were further complicated by intense internal power struggles among the ghair-mulkis and by British efforts to consolidate their own influence. After a particularly nasty interaction in February 1897, Agha Mirza determined that he could no longer continue to weather the city's intrigues.[21] Though he still had the support of the nizam, he requested permission to leave. When it was granted, he immediately boarded a train and left Hyderabad for good, without even pausing to see his wife or children. Such was Agha Mirza's influence in the city that his departure marked "a turning point in the life of Hyderabad politics."[22]

His family, however, stayed behind. Already intimate with the nizam, his children and particularly his son-in-law Hamidullah Khan were deeply enmeshed in Hyderabad's upper echelon. Thus, rather than leave with her father, Begum Sarbuland instead made frequent trips to visit him at his homes in Lucknow and Shimla. Even from a distance, Agha Mirza continued to have a presence in her life. When he left in 1897, he was less than fifty years old. His name appears frequently in his daughter's 1909–10 travelogue, as she writes him letters, prays for him, and notes the gifts that she bought him. Though he lived for nearly thirty-five more years after leaving Hyderabad, his break with the city was complete. Even in 1910, her parents seemed to be living separately, her mother in Hyderabad and her father in Shimla. Agha Mirza finally passed away in 1933 and was buried in Delhi, a few feet from the tomb of the famed sixteenth-century Sufi saint Khwaja Baqi Billah, near the gates of Delhi's old city. His daughter would join him in a nearby plot a quarter century later.

Begum Sarbuland's Childhood Education and Early Life

In the late nineteenth century, most Indian women were given no formal education beyond basic religious instruction. Fewer than 1 percent of women could read. Those who could also write were even more scarce. Writing was considered more dangerous than reading, since it would allow women to communicate with unrelated men and thus endanger the family's honor.[23] As one woman who grew up in the 1850s reflected: "It had long been customary in my family to teach the girls how to read—teaching them how to write, however, was strictly forbidden. The girls were taught only to vocalize the Arabic of the Quran and read a bit of Urdu so that they could gain some knowledge of their faith and the rules of prayer and

Fig. 11.2 A "code of ethics for travelers" written by Begum Sarbuland. *Source: Shakira Sultana.*

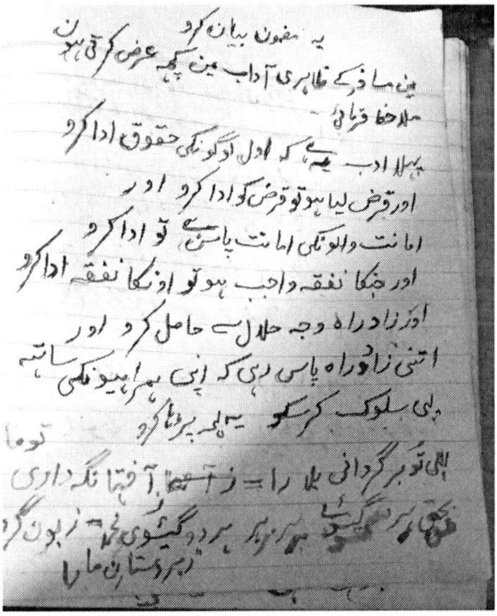

fasting. . . . I was, however, still dying to learn to read Urdu, but I couldn't find any woman who could teach me."[24] In this regard, Begum Sarbuland's family was liberal: her grandmother Munavvar Zamani Begum could read and had enough knowledge of Urdu and Arabic to be able to give her son his early religious instruction. This made her exceptionally well educated compared to most Indian women.[25]

The situation regarding women's education had improved somewhat by Begum Sarbuland's time, but even in the 1870s, there were almost no opportunities for girls to attend a formal school.[26] Some schools had been founded in Hyderabad, largely by Christian missionaries or by individual families for their own daughters.[27] Begum Sarbuland was instead educated at home, probably by her mother and visiting instructors. She was certainly given a rigorous religious education, learning to read the Qur'an and read and write in Urdu. She could understand written Persian, as indicated by her occasional reference to Persian poetry in her private notebooks, but was probably never able to speak it fluently (though her male relatives certainly could). As a youth, she learned music, sewing, embroidery, and board games, such as chess.

Her travel diary, of course, demonstrates not just that she had cultivated writing as a lifelong habit but that she was in fact an excellent and even eloquent writer. Some of her writing is preserved in private notebooks and diaries. The entries include personal memories, activities, important life events, and occasional compositions. (By way of example, figure 11.2 shows a code of ethics for travelers written in her own hand.)

Her diary also mentions a long list of correspondences she kept, demonstrating that she was a prolific writer of letters. She also likely subscribed to and read Urdu women's magazines, which thrived from the early twentieth century.[28] Women's writing in Hyderabad began to flourish from the early twentieth century, and while not everyone published, Begum Sarbuland herself was an active participant in this culture of writing.[29] She no doubt participated in oral literary gatherings, perhaps as a member of one of the many "purdah clubs" for women that were popular at the time.[30] In addition to her prose and letter writing, Begum Sarbuland was also an avid poet. She used the pen name "Akhtar." She references writing her own poetry twice in her diary. While meeting with some women in Constantinople, she sings "one of my own ghazals." When she meets with the Prince and Princess of Wales, it is revealed that she had written a poem in Urdu to commemorate their visit to Hyderabad. Somewhat surprisingly, then, there is very little poetry in her diary, neither verses of her own nor by other poets. Prose writing in Urdu is often filled with verse quotations, but her travelogue includes almost none.[31] Despite this, she loved writing poetry, and much of her verse remains preserved in her private notebooks.[32] (See a draft of a ghazal poem written in her own hand in figure 11.3.)

Begum Sarbuland's access to education seemingly expanded after her marriage in 1894. Among the new class of English-educated Indians, the group to which Hamidullah belonged, wives were increasingly expected to be educated life partners who could share in their spouse's intellectual interests. Hamidullah perhaps went further than many of his peers in this, being himself strongly in favor of reforming and expanding Muslim women's education. At some point after their marriage, then, Hamidullah likely encouraged Begum Sarbuland to continue her education, particularly by learning English (or, at the very least, he did not prevent her from doing so). Her diary suggests that she had learned English relatively recently but that, by 1909, she was fluent enough to read English novels, travel alone, and befriend a range of British women. She was even able to banter comfortably with the Princess of Wales. (When the royal couple praised India's beauty to her, Begum Sarbuland quipped, "It belongs to you.")

It was in the 1900s that she first accessed formal education, in a way. By then a few prominent schools had opened to educate wealthy, sharif Muslim girls like Begum Sarbuland's daughters. In Hyderabad, the premier institution of this kind was Mahbubiya Girls' School (originally called the

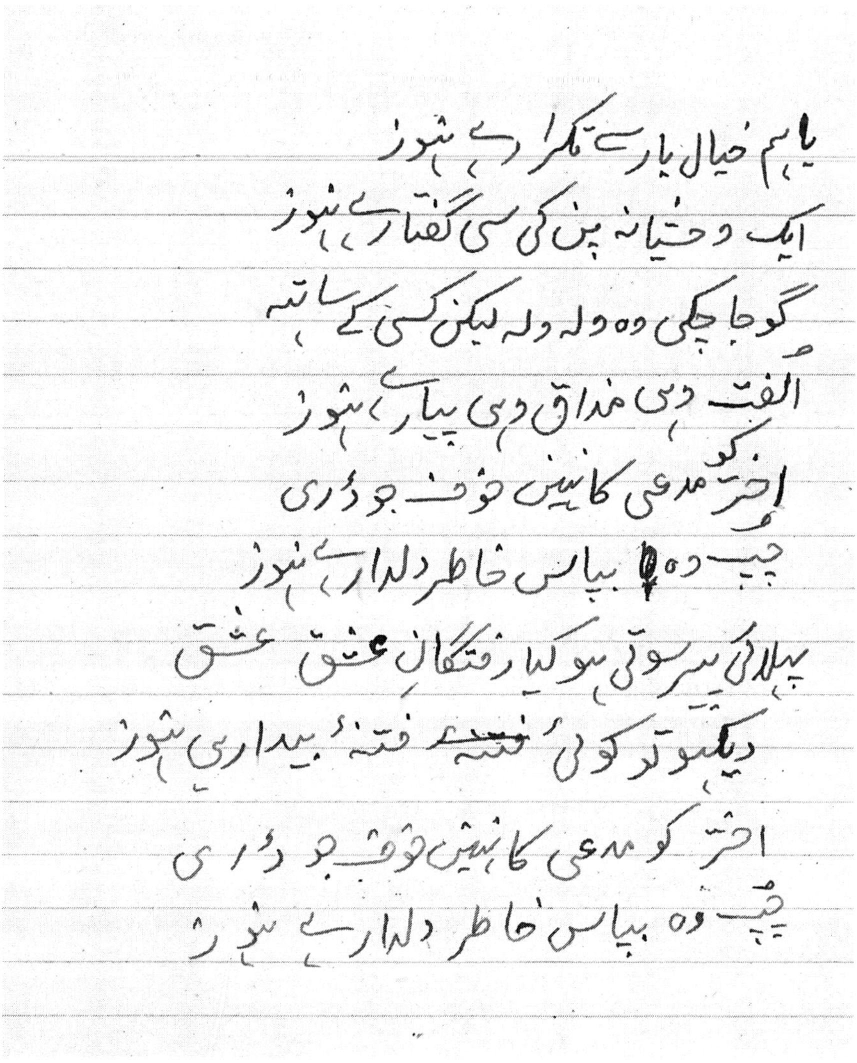

Fig. 11.3 Poetry taken from the private notebook of Begum Sarbuland.
Source: Shakira Sultana.

New Zenana School), which opened in 1907 as a "finishing school for girls in the upper echelons of society, which prepared its students for the Cambridge local examinations and for their future role as wives and mothers."[33] Begum Sarbuland's daughter Khurshid Begum was promptly enrolled. The initial class included twenty-two students, including the daughters of Sarojni Naidu and Lady Amina Hydari.[34] Not enrolled but attending as an auditor was Begum Sarbuland, who insisted on sitting in on her daughter's classes so that she could learn herself! In 1909 she further supplemented her education by studying colloquial Arabic, which she learned to at least a passable level.

Married Life and After

When Hamidullah Khan married Akhtar un-Nisa, she was about eighteen, and he was just shy of thirty. They had some similarities and some differences. Both descended from Mughal families from Delhi.[35] However, while she was the inheritor of India's precolonial princely traditions, the living embodiment of life among India's princely elite, Hamidullah was a part of the new colonial avant-garde. He represented the "new elite," who were trained in England, comfortable in English, and worked closely with the British administration.

The Life of Hamidullah Khan

Hamidullah Khan was born in Agra on March 17, 1864. Almost nothing about his mother has been recorded, except that she was the daughter of Nawab Ikhtiyar ud-Daula Khwaja Ali Ahmad Khan, whose own family descended from the Mughal court.[36] His father, on the other hand, attained significant fame, and his family history is well-documented.[37] Samiullah Khan (1834–1908) was an Arabic scholar by training. Widely respected by northern Indian Muslims, he was held in great esteem by the British, with whom he worked for many years as a court judge. In colonial British sources, he is best remembered for his services in solidifying colonial control over Egypt in 1886.[38] His contributions there earned him a knighthood. Today, though, he is known primarily for advocating educational reform among Indian Muslims.

Hamidullah's education was rigorous and thorough. In Agra, he studied with accomplished scholars from almost the moment he could walk. In 1871, the family moved to Allahabad, where, from age seven, a host of tutors schooled him in Urdu, Arabic, Persian, and English.[39] Within two years he had apparently so sufficiently mastered Urdu and Persian that his curriculum "was confined to English and Arabic" alone.[40] In 1873, a judicial posting took the family to Aligarh. Around this time, Hamidullah joined his father and the famed scholar Sayyid Ahmad Khan (1817–1898) on a tour of Punjab to drum up support for a new school for Muslims in Aligarh that would teach both Western and Islamic subjects. Hamidullah's role on this trip was to demonstrate the potential of such an education.[41]

Officially opened in 1875, the Mohammadan Anglo-Oriental College was not just a school but the epicenter of an entire reformist movement. The college was the brainchild of Sayyid Ahmad Khan, but Samiullah Khan was among its earliest and fiercest advocates. Like Sayyid Ahmad, he believed that Indian Muslims could only return to their former glory by reforming their society and embracing Western knowledge, particularly among the *ashraf*, or upper class. The college was intended to provide this transformative, composite education. Despite intense criticism and even accusations of blasphemy, the school was a stunning success: even today the word "Aligarh" serves as a shorthand for a certain intellectual orientation and the lifestyles and careers of its alumni. Hamidullah made history by becoming the school's first student and one of its most successful graduates.[42]

Between the school's rigorous education and his early training, Hamidullah advanced rapidly. By age fourteen he was ready to sit for the university entrance exam.[43] He was offered a two-year scholarship by the maharaja of Patiala, which he took up in 1879, continuing his education from Aligarh. He also began private tuitions in Latin, classical English literature, and classical Arabic. By this time, his father had been transferred to Muradabad, but Hamidullah remained in Aligarh with Sayyid Ahmad Khan. His diary entries, many of which were later published, give a picture of his life during this period.[44]

The second phase of his life began in 1880. When he was sixteen, Hamidullah moved to England.[45] After two years of rigorous study, he was admitted to Lincoln's Inn and Christ's College, Cambridge. He graduated in 1886, was called to the bar, and worked for the India Office in London.[46] He made several long trips during this time, including one to Ireland and another to Vienna, where he represented the India Office at a major Orientalist conference.[47] He also traveled frequently to France to learn French, but the most significant voyage he took was to Egypt in 1884. There, his father was tasked with reviewing the courts and convincing Egyptians that British rule was compatible with Muslim law. Despite being twenty years old, Hamidullah was invited to join the mission as a personal assistant to its leader, the Earl of Northbrook. This appointment made him the third generation of his family to serve Britain's imperial expansion.[48] After returning to India in late 1886, Hamidullah pursued a career in law, moving to Allahabad to work as a lawyer in the high court. Over the course of the next decade, he grew his legal practice and edited a major intellectual and literary magazine, the *Allahabad Review* (founded 1890). He also founded an organization dedicated to Muslim educational reform and published translations and commentaries in English on the Persian poetry of Hafiz.[49]

Hamidullah Khan Weds Akhtar un-Nisa

Hamidullah and Akhtar un-Nisa were married on January 14th, 1894.[50] The grandeur of the wedding can be seen through a travelogue written by a wedding guest, Shah Muhammad Danapuri. Danapuri often uses the present tense, as though he were narrating the events as they occured: "The guests are beginning to arrive and the crowd grows larger and larger at every moment. The stream of guests flows ceaselessly. One can see many very fine people here; it is a beautiful gathering. Many large houses have been arranged and prepared for the guests. [As we enter the room arranged for us,] two attendants, both residents of Delhi, come before us with their hands folded. They appear to be diligent and courteous. Maulvi sahib [Samiullah Khan] has assigned two such servants to each guest in attendance."[51] The guests were provided with every necessity. Food and beverages were always available. Delhi can be frigid in January, and Danapuri was delighted to find that warm water was left outside his room every morning before dawn for his ablutions.

In his entry from January 14, 1894, Danapuri describes the groom's procession in detail:

> It is the seventh of the month of Rajab, the day selected for the marriage ceremony. . . . Before us there is a long line of phaetons and carriages. The hosts are busy inviting their guests to sit in them. A little later the bridegroom, Muhammad Hamidullah Khan, emerged from the house dressed in a wedding costume. He was wearing an Arab cloak in the Medina style that had been specially made in Mecca, and a white turban made of Syrian cloth. He rode an iron-grey horse, and his elders walked beside him. . . . Setting off, the wedding procession passed the Jami Mosque and continued toward Nawab Sarvar Jung Bahadur's [Agha Mirza's] house in the Churiwalan neighborhood. The distance was more than a mile, but everyone was in high spirits. They were even more joyous than they had been during the *sachaq* procession [the day before]. [At the ceremony] Janab Maulvi Samiullah Khan Bahadur conducted the wedding himself and sought the assent and agreement of both parties [for the marriage]. Cries of *mubarak*! and *salamat*! [congratulations] rang out.[52]

Then sweets were distributed to the guests. The proportions were massive, with each guest receiving over three pounds of sugary goodness in large china bowls. "Now everyone who had attended the marriage stood up and went outside to pray. There we saw the items that had been included in the dowry. Truly, Nawab Sarvar Jung [Agha Mirza] has shown extreme generosity during this wedding."[53] The festivities concluded the following day with a meal, the *valima*, hosted by Hamidullah's family.

After the wedding, Hamidullah stayed in northern India, working and attending to his ailing mother. Begum Sarbuland may have stayed there with him for a time before returning to Hyderabad. After his mother's death in September 1894, he closed his practice in Allahabad and moved to Hyderabad. He arrived there in 1895. Begum Sarbuland gave birth to her first son, Muhammadullah, and then a daughter, Khurshid Begum (1896–1981), the following year.

Married Life in Hyderabad

Hamidullah was highly qualified, but it was his marriage to Akhtar un-Nisa that earned him an immediate position among Hyderabad's ruling class. After all, his father-in-law was the second most powerful man in the kingdom. Hamidullah's first post was as a judge of the high court (plates 3 and 5). The nizam also granted him two honorifics: Afzal al-Ulama (Most excellent of scholars) and Nawab Sarbuland Jung (Exalted, lit. "having the head held high"). These titles are said, perhaps apocryphally, to have been arranged by his wife. Family lore tells that Akhtar un-Nisa wished for her husband to receive a British knighthood and that she repeatedly pressured the nizam to raise the matter with the British resident directly. If this had happened, he would have become Sir Hamidullah (and because she used her husband's title as her name, she would have been "knighted," in a sense, herself). The nizam, though, was unwilling or unable to make this happen. Instead, he tried to pacify her by offering her husband a clever homophone. Rather than make Hamidullah a "sir," he made him a "sar" by granting him the title Sar-buland ("exalted"). From then on, he was known as Nawab Sarbuland Jung, and his wife became Begum Sarbuland Jung (the wife of Sarbuland Jung).[54]

Hamidullah's arrival in Hyderabad changed more than Akhtar un-Nisa's name. By the end of 1896 they already had two children. The size of their family then grew almost annually. They reportedly had a total of thirteen children, of which at least ten survived to adulthood. These were Muhammadullah (b. 1896),[55] Khurshid Begum (1896–1981),[56] Mahmudullah (b. 1901), Masudullah,[57] Maqsudullah, Qudsiya Begum, Dr. Wajihullah Alavi,[58] Halimullah (1909–73),[59] Tehzib Nur Begum (1912–96),[60] and Usman Ilahi Bakhsh (b. 1914). Hamidullah's career advanced quickly too: in 1904, he was promoted to home secretary, and four years later he became

Fig. 11.4 A photograph of Nawab Hamidullah Khan Sarbuland Jang, circa 1907–8. *Source: Daniel Majchrowicz.*

the chief justice of the supreme court (fig. 11.4 and plate 6). At forty-four years old, he had a salary of two thousand rupees per month.[61] He did not retain the position for long, however.[62] The following year he took a leave of absence to travel to Europe and the Middle East with his wife. In 1912, he claimed a full pension from the state and took an early retirement.[63]

Prior to his retirement, though, the couple made at least one journey to Delhi, in 1911, to attend the Imperial Durbar, where Hamidullah was sent as a member of the Hyderabad delegation.[64] The Delhi Durbar of 1911 was a lavish commemoration of the coronation of King George V and Queen Mary (whom the Sarbulands had visited in London before they were coronated) as emperor and empress of India. Unsurprisingly, Begum Sarbuland's attendance is not recorded in the official proceedings. Nevertheless, stories passed down by her children confirm that she did attend and that she may have done so assertively, all while wearing a blazing red outfit she had designed and sewn herself (plate 29): "Upon her return from her travels, [Begum Sarbuland] created two blouses from her wedding dupatta, one of which she wore to the 'Durbar' [of 1911]. . . . Dressed as befitting a Mughal princess and the wife of the Chief Justice of the Hyderabad Court, she walked straight up to the dais reserved for 'dignitaries' that is, the British women, ignoring all protests from the guards. Her self-assurance intrigued the Viceroy's wife who then wanted to meet her. There followed an interesting exchange and conversation in English as they sat together."[65] While there is no written record of this account, this oral narrative speaks both to Begum Sarbuland's forthrightness and knack for making friends on every possible occasion. And of course, her acumen for fashion and clothing design.

After Retirement

Hamidullah's retirement brought massive changes to Begum Sarbuland's life. He had once traveled to Hyderabad for her; it was now time for her to move to Allahabad for him. They shifted there in 1912, moving into a house in the city's Civil Lines. The reasons behind the move are unclear, but it may have had to do with his health or the pressures of his job. In Allahabad, he reopened his legal practice. It was she who primarily took charge of the family's affairs in Allahabad, likely devoting much of her attention to her children (plate 8). As their children grew older, the family attended or organized a series of weddings, including an extravagant event for their eldest son, Muhammadullah, in 1916 (plate 10).[66] At the same time, the couple began to dedicate themselves more than ever to the cause of

Muslim educational reform. Hamidullah paid special attention to his alma mater, Aligarh Muslim University (as the school was called by then), which he served as a trustee, but he also headed or attended a range of educational conferences. As he aged, though, he gradually curtailed his activities and withdrew from society.[67] He spent the later years of his life editing and publishing his writings, including his earlier travel diaries.

His wife, meanwhile, became ever more active. She continued to travel to Hyderabad regularly (plate 9). She joined various Muslim women's reformist organizations, traveling across India, probably without her husband, to attend their conferences and events. Her name appears consistently in meeting minutes and conference bulletins over a period of several decades. One instance comes from 1918, when she represented Allahabad at the first meeting of the All-India Ladies Association in Bhopal.[68] She attended the conference with her daughter Khurshid Begum, who was the representative from Aligarh (plates 13 and 16).[69] At the conference, Begum Sarbuland blamed infant mortality in India on "parental ignorance" of the proper manner for feeding and caring for a child. She proposed a resolution that would call on the government to increase the quantity and training of "lady doctors" and recommended competent nurses pay home visits to ensure that parents were properly informed of the best child-rearing practices.[70] On another occasion, in 1927, she gave a speech at the All-India Women's Conference on Educational Reform in Pune.[71] She probably also attended the same conference again in 1928 and was selected there to join a delegation on child marriage that was sent to meet with the viceroy.[72]

These events were more than just professional affairs. They were likely a key part of Begum Sarbuland's social life. In Hyderabad, she had always kept a wide circle of female friends. These conferences provided her the opportunity to meet with her friends and socialize with like-minded women from across India. A travel diary entry by the Hyderabadi reformist Sughra Humayun Mirza, who also attended the 1918 conference in Bhopal, gives a sense of this sociality. She writes: "At eight o' clock we all went to the dining hall, that is: 1. Nafis Dulhan Sahiba, the secretary of the All-India Muslim Ladies Conference; 2. Abru Begum Sahiba, the secretary of the Ladies Club; 3. Fatima Arzu Begum, the secretary of the All-India Ladies Conference; 4. Begum Sarbuland Jang; 5. Mrs. Abd ul-Majid [Begum Sarbuland's daughter Khurshid Begum]; . . . We all had at dinner together. After the meal we played the harmonium and sang songs. This lively gathering lasted until eleven o'clock."[73] As her children grew older and as her husband's health continued to deteriorate, Begum Sarbuland seems to have found increasingly more time for these reformist events.

She also spent a significant amount of time visiting her family, including particularly her father, who passed away in 1933 (plates 12 and 19). She made frequent visits to Aligarh, where her eldest daughter, Khurshid Begum, lived. She seems to have found plenty of other reasons for travel too. One trip from 1923 is recorded in a travelogue written by her sister-in-law Rahil Begum Shervania (1894–1983). Rahil Begum had married Begum Sarbuland's brother Yahya Beg in 1920 but was widowed three short years later. To console herself for the loss of her husband, Rahil resolved to perform the hajj that same year. This was a challenging journey for a newly widowed woman with two young children, but Begum Sarbuland was eager to help and give her advice on how to prepare for the journey. When she learned about Rahil Begum's intention to perform the hajj, Begum Sarbuland traveled to Lucknow and brought her to Allahabad for a week.[74]

In 1925, Begum Sarbuland and Hamidullah moved to Delhi. By then, her husband was ailing so much that in 1929 he was forced to skip his own son's wedding (plate 15). They began to spend their summers in the mountain towns of Shimla and Solan and their winters in Delhi.[75] He passed away in 1930 at the age of sixty-six, probably of cancer.[76] Begum Sarbuland describes his death as a terrible blow.[77] He had been an immense force in her life, and the pair were very close. The nature of their relationship that emerges from her travel diary is one of concord and mutual support. They appear to take decisions together, while allowing the other to explore their own interests freely. Hamidullah was nominally the head of the family, but in truth, Begum Sarbuland lived an independent life both inside and outside the home. Though their marriage had been arranged, it was an ideal match. The two shared many qualities and were similarly spiritually aligned. They were deeply religious and committed to the defense of Islam. Some of the most poignant moments in Begum Sarbuland's diary are when they worship together at sacred sites in the Middle East. She had an intellectual bent and a deep desire to observe and to learn. Hamidullah was of course a scholar who had trained with the greatest thinkers in Britain and India. He clearly appreciated his wife's intellect and curiosity and sought to cultivate these qualities.

More broadly, Hamidullah seems to have always been a particularly liberal and freethinking individual. He was an early supporter of women's suffrage; in the 1880s, he even aspired to become a vegetarian.[78] He avoided alcohol but drank it occasionally on the principle that it was better to follow local custom than to cause disruption or offense.[79] In 1886, he lamented Britain's insistence on carting off Egyptian antiquities to store away in London's museums.[80] Though he critiqued some aspects of colonialism, neither husband nor wife ever seem to have objected to British rule in India. They involved themselves in a range of reformist affairs, but there is no record of them having taken part in India's surging anticolonial movements in which even some of their children were deeply involved, whether with the Indian National Congress or the All-India Muslim League. They were perhaps too deeply invested in British India to advocate for its demise, or perhaps they simply wished to dedicate their time to societal issues rather than political causes. By the same token, there is also no indication that he or his wife ever supported the idea of Muslim separatism within or from India.

Life after Hamidullah's Death

Begum Sarbuland lived a quarter century longer than her husband. Though her diary states that she was "deadened to the pleasures of life" after her husband's passing, widowhood did not dampen her vigor in the long run. She maintained her vibrant social life, and given that a significant number of high-ranking figures were among her friends and acquaintances, a group that included even the nizam himself, she retained her political clout (plate 17). In the 1920s and 1930s, she frequently met with the wives of India's viceroys, who were the most influential women in the British Indian hierarchy, including Lady Irwin, Lady Reading, and Lady Willingdon.[81] These visits were likely connected to her various reformist and charitable activities related to women's welfare, but the fact that she was so regularly given audiences attests to her enduring influence.

In the 1930s, she expanded her advocacy work in surprising new directions. On April 20, 1936, for example, she gave a broadcast speech on the topic of dowry. The speech lasted approximately fifteen minutes. Perhaps to appease readers after a pedantic lecture, her segment was bookended by two soft musical numbers: P. S. Mukherjee playing the sarod and some ghazals sung by Muhammad Shafi.[82] The most dramatic step she took during this period was the publication of her travel diary. The volume is not dated, but it was likely published around 1935–36 (plate 21).[83] Begum Sarbuland understood her diary to be a part of her reformist work. She had long wished to open a vocational school for indigent Muslim girls.[84] She cast this book as a part of that project. A note on the back page promised that all proceeds from the sale of the book would be applied to the construction of such a school in Delhi (see appendix 1). That particular project did not come to fruition, but in 1942 she did manage to open a school in Hyderabad for orphaned and indigent children, called Tarbiyat Gah.[85] She may have also run similar schools in Delhi and Allahabad, the later perhaps with her daughter Khurshid (plate 20).

At some point, perhaps in the early 1940s, she seems to have moved to Hyderabad, where she still had family and a social circle. Her surviving grandchildren today recall living with her in the Lal Tekri (or Red Hills) area of Hyderabad in the early 1940s. She lived in a large house with the capacity to host the children and any guests who visited her. At home, she wrote poetry and played board games, chess and pachisi particularly. Her steadying presence seems to have been particularly important to the family during World War II. Between 1942 and 1945, the husband of her daughter Usman Ilahi was taken by the Japanese as a prisoner of war in Singapore.[86] Rather than stay alone in Lahore, his family moved to Hyderabad, where they joined other members of Begum Sarbuland's family who were already there. Her grandchildren recall that she continued to maintain close links to the royal family. Every few months she would be sent large baskets of sweets and foodstuffs.[87] On some occasions, Begum Sarbuland herself would visit the royal household at Nazri Bagh, or King Kothi Palace, in Hyderabad.[88] Even late into her life, then, Begum Sarbuland remained an influential and well-connected figure surrounded by her large family.

Begum Sarbuland traveled frequently, often accompanied by either her daughter Khurshid or her granddaughter Akhtar (plate 22).[89] These trips were, again, highly social affairs. In March 1934 she went on hajj, probably for the second time. The evidence for this journey comes from a travelogue from that year by the newspaper editor and activist Fatima Begum. On the ship to Jeddah, Begum Sarbuland befriended Fatima Begum and the famed poet and activist Hasrat Mohani and his wife, the anticolonial firebrand Nishat un-Nisa Begum Hasrat Mohani.[90] Fatima Begum records that Begum Sarbuland took a leading role in organizing life at sea: "At Begum Sarbuland's suggestion, a religious gathering and sermon was arranged for the female passengers."[91]

What was probably her third pilgrimage occurred in 1937.[92] Records from the British consulate in Jeddah record her presence in the city in July of that year:

> Lady Abdul Qadir . . . arrived at Jedda with one son on the 28th of July.[93] She is accompanied by Begum Sarbuland Jung (widow of a high personage in the Government of Hyderabad), her son and granddaughter,[94] as well as an Englishwoman named Miss M. M. Clements.[95] The latter says she became a Moslem in England thirteen years ago, and hopes to visit Medina and Mecca with the Jung party, finally going to India with them. . . . The Jung party found that a day and a half of Jedda in July was enough for them, and left for Medina with Lady Qadir and her son, leaving the unfortunate Miss Clements to wait here for her Islamic "bill of health."[96]

The broader context for this journey was her son Halimullah's health (plate 18). While studying medicine in Edinburgh, he spat blood, indicating tuberculosis. Doctors advised him to convalesce by the sea, and so he moved to Nice, France. His mother joined him there in early 1937. On the way home they visited Medina and probably Mecca as well.

The Jung family archive contains a striking photo of Begum Sarbuland in Nice (plate 23), probably taken in the late spring of 1937. In it, she is standing between her granddaughter Akhtar Sultan Begum and the exiled Ottoman sultan and deposed caliph, Abdülmecid II (1868–1944). The group is all smiles. In the picture, Begum Sarbuland is wearing a sari with a full-sleeved blouse and a cardigan. The tail of her sari is draped over her head, covering most of her hair, and looped back around over her chest. She stands close to Akhtar, who is a few months pregnant with a son, Mansoor Hasan. Both women are at a remove from the former caliph, a dapper older gentleman dressed in a suit, tie, and fez. His hands rest lightly in his trouser pockets. Combined with the women's smiles, the effect is one of casual ease, of a quick snap of a few friends on a pleasant spring afternoon in the south of France. There is no doubt that the begum and the caliph would have had plenty to talk about, for both were intimately tied to the Hyderabadi royal family. Her connection to Hyderabad is obvious, but his is remarkable. After the toppling of the Ottoman Empire and the abolition of the caliphate two years later, Abdülmecid was exiled. He fled for Switzerland.[97] A few months later he moved to Nice. Turkey denied him financial support in his exile, but he found succor from Hyderabad's ruler, Mir Osman Ali Khan (1886–1967), who proposed to unite their families through marriage. In 1931, the nizam's first son, Azam Jah, was married to the caliph's daughter Dürrüşehvar (1914–2006), while his second son, Muazzam Jah, was married to the caliph's niece Nilüfer (1916–1989). Both women, still in their midteens, promptly sailed for Hyderabad. They quickly learned to wear saris and speak Urdu while at sea and arrived ready to join Hyderabad's social and political life. Begum Sarbuland had surely met these women on one of her many visits to the city, and now she met their father in Nice.

Later Life

At some point, perhaps in 1946 or early 1947, Begum Sarbuland seems to have moved from Hyderabad to Delhi, where she stayed with her youngest son, Halimullah, and his family in the house built by her late husband in Daryaganj (plates 24, 25, and 26).[98] When India gained its independence from Britain and was partitioned to form India and Pakistan, Delhi was engulfed in violence, with many of the atrocities taking place in the old city near where Begum Sarbuland was then living in her home, Hamid Manzil. Her granddaughter Shakira Sultana, who was there at the time, recalls that while they were in hiding, the power was cut, and an eerie silence prevailed. In the darkness they could plainly see a raging fire burning in the direction of the Jama Mosque in Old Delhi. They witnessed several murders. One can imagine that Begum Sarbuland might, in these moments, have recalled the stories her father would have told her about his own experiences fleeing Delhi during the uprising of 1857.

In search of safety, the family took shelter in one of Halimullah's restaurants in the formerly British-dominated area of the city. They then rented an apartment in Connaught Place, which was deemed to be safer than returning to Daryaganj. While they were away, though, their home was declared evacuee property by the state and occupied by a refugee from Pakistan.[99] The family contested the confiscation of their home in court and through their contacts, though it was not returned to them for several years. During some of this time, Begum Sarbuland likely lived in Hyderabad, but she spent the final years before her death in Delhi, living with Halimullah and his family. She was also in regular contact with her other children. Even her daughter Usman Bakhsh Ilahi, who was living in Pakistan, came from Lahore to visit her yearly until her death, despite the complications and difficulties involved in crossing the border at that time.[100]

Over time, Begum Sarbuland developed dementia. She increasingly lost her ability to recognize her family members, particularly those who were not her children, but retained a playful and energetic spirit. Her grandchildren recall her frequently claiming to have befriended a jinn who would provide gifts for them; the children would then be sent to search the house to find out where the jinn had hidden the presents. However, Begum Sarbuland increasingly required assistance to survive. Throughout her life, she had relied on her personal servants, Hindu and Muslim, many of whom spent their lives in her service. In her final years, she was increasingly supported by Bakhtavar, an orphan who joined the family to assist her (plate 27). By the end of her life, Begum Sarbuland was spending six months of the year in the family's vacation home in Solan (plate 28). In 1957, she was in a car driven by her son Halimullah, going from Solan back to Delhi, when the heat and humidity became too much for her. She reportedly passed away soon after arriving in Delhi.[101] She was buried in an unmarked grave near her husband and father beside the shrine of Khwaja Baqi Billah (plate 30).[102] Begum Sarbuland was survived by a huge number of grandchildren, many of whom went on to become successful and well-known figures in India, Pakistan, and the United States. In part, the success of her descendants can be credited to her active involvement in crafting the family's legacy. Today, nearly seven decades after her passing, she remains a legend in the family.[103]

Conclusion

Begum Sarbuland lived a long and remarkable life at the heart of India's political and cultural worlds. Though she declined to cultivate a public persona, she was not demure and certainly did not live a quiet life behind the walls of her home. Those around her knew that she was a force to be reckoned with, quick-witted and confident. She was a woman with many friends and impressive influence. She had a bold and daring personality. She felt little shyness about approaching strangers and asserting her will. In her diary, she is constantly finding new companions and making new friends; these were apparently lifelong traits. Still, in all her interactions she was careful to maintain what she considered the appropriate Islamic levels of modesty. In her younger days she wore a full veil but reduced its use before disposing of it entirely. Photos taken of her that date from 1910 to 1957 reflect this gradual shift. This transition can be seen in the images included in this book.

Despite her dynamic approach to purdah and her passion for fashion, Begum Sarbuland dressed modestly until the end of her life, preferring long-sleeved blouses paired with a sari, the loose end of which was always draped over her head, as in the photograph of her with the caliph. Begum Sarbuland believed in purdah as an ideal Islamic practice. She reflected on it often and shared her considerations in her diary. For her, purdah did not mean staying at home; nor did it mean always remaining completely covered. She adapted her purdah according to the circumstances, but she never disposed of the concept entirely. Even in her final years, she was fond of saying that when women began to go about with no purdah, the end times (*qiyamat*) would be near.[104]

Begum Sarbuland was above all a devout Muslim and a Sufi. She took guidance from spiritual guides and frequented the shrines of saints. She was initiated into the Qadiri and Chishti Sufi orders, though she was most dedicated to the latter. Her daily life was shaped by her faith, and it was her faith that shaped how she saw the world around her. She was a fervent and passionate devotee of the Prophet. She often said that after her death, she would not remain in her grave but would follow a tunnel straight to Medina and that if anyone wished to give her a salaam, they should not visit her grave in Delhi but speak their greetings in the Prophet's city.[105] Perhaps she is there now.

Notes

1. Interview with Shakira Sultana, May 27, 2022.

2. Not every woman in her society did the same. One of her contemporaries was Begum Sughra Humayun Mirza (1884–1958), another Hyderabad-born woman who married a lawyer from northern India. Begum Sughra Humayun Mirza was similarly committed to reformist movements. The two women differed, though, in their attitude toward public engagement. Unlike Begum Sarbuland, Sughra published and lectured frequently, generating a wide following and significant fame. For a brief biography and bibliography, see Majchrowicz, "Sughra Humayun Mirza," 451–54.

3. They were never confined to the city, though. In the late eighteenth century, her great-grandfather Mirza Akbar Beg Khan performed the hajj in Mecca, visited Egypt, and reportedly even studied mathematics in Italy. He also traveled to southern India and spent a few months in Hyderabad. (It is unknown whether Chhoti Begum, his wife, joined him or stayed in Delhi.) Agha Mirza Beg, *Karnama-e Sarvari*, 19. Begum Sarbuland's grandfather and father traveled widely too, but, like Akbar Beg, they also always returned to Delhi. The family remains closely associated with Delhi today: in the 2010s Begum Sarbuland's grandson Najeeb Jung was the city's lieutenant governor.

4. For an accessible look at the functioning of the British residency system in Hyderabad, see Dalrymple, *White Mughals*. For an India-wide examination of the institution, see Fisher, *Indirect Rule*; Keen, *Princely India*; Ramusack, *Indian Princes*.

5. Ghalib was his paternal grandmother's brother and his uncle-in-law, since Agha Mirza's mother-in-law, Aziz un-Nisa Begum, was the daughter of Ghalib's brother Yusuf Mirza. She had a close relationship with the poet. Agha Mirza Beg, *Karnama-e Sarvari*, 12.

6. Munavvar Zamani Begum's parents were Qarshiyah Begum and Khalilullah Khan. Qarshiyah Begum's parents were Shahzadah Begum and Mirza Bakhtavar Bakht. The latter was a grandson of Shah Alam II. Agha Mirza Beg, 2.

7. Agha Mirza Beg, 13. She was also known as Makhole Begum.

8. See Minault, "Master Ramchandra."

9. Server-ul-Mulk, *My Life*, 48–53.

10. Oldenburg, *Colonial Lucknow*, 237.

11. Campbell, *Glimpses*, 365; Server-ul-Mulk, *My Life*, 32.

12. His autobiography includes a fascinating narration of the long journey south. He departed from Lucknow in May 1872 and arrived in early 1873. Agha Mirza Beg, *Karnama-e Sarvari*, 61–81. See also Server-ul-Mulk, *My Life*, 79.

13. Among his additional pupils were Saadat Ali Khan (b. 1864), Munir ul-Mulk, and Mir Laiq Ali Khan, Salar Jung II (1861–89). Server-ul-Mulk, *My Life*, 119. The latter would go on to become prime minister of Hyderabad and was himself the author of a Persian-language travelogue. See Laiq Ali Khan, Sharma, and Khalidi, *Vaqa'i-yi musafarat*. He also taught Maharaja Kishen Pershad, who would twice become prime minister in the twentieth century and was a close contact of Begum Sarbuland's husband.

14. A tutor, Captain Clerk, who did not speak Urdu, was brought from England to supervise his education. Agha Mirza was appointed as his assistant, and together, they tutored the young nizam, largely in English. They were briefly joined by the famed Urdu writer Deputy Nazir Ahmad, whom Agha Mirza did not like and who apparently tried to get Agha Mirza fired. Server-ul-Mulk, *My Life*, 155. Interestingly, the British resident and his tutor were constantly trying to get the nizam to travel. In 1875 the British resident insisted repeatedly that the young nizam make an educational voyage to Bombay. This was resisted heavily by the court and never happened. Server-ul-Mulk, 145. He instead took educational travels in his own cities of Gulbarga and Aurangabad, again at the instigation of his British minders. Server-ul-Mulk, 169–77. Then, at age eleven, he traveled to Delhi to attend the coronation durbar of Queen Victoria. Agha Mirza Beg, *Karnama-e Sarvari*, 194–200. Soon after, the British resident apparently tried to trick the court into having the nizam travel to England. Agha Mirza Beg, 201–2. That effort failed, but he later went to Calcutta as a capstone for his education and to mark his assumption of his full powers. Agha Mirza Beg, 222–30. Such efforts to force Indian rulers to travel were common at the time. See Majchrowicz, *World in Words*, 84, 118.

15. At this time, Agha Mirza was granted the title Sarvar Jung Bahadur and given a sizeable pension that was guaranteed to him and his offspring in perpetuity. He was later granted further titles, including Sarvar ul-Mulk. Campbell, *Glimpses*, 358.

16. Cohen, *Appeal*, 257; Server-ul-Mulk, *My Life*, v.

17. His wife would have been about sixteen, and he was about twenty-six. Server-ul-Mulk, 121.

18. These were Maryam Sultan, Zulqadar Jung, Yasin Sultan, Usman Nawaz Jung, Khursheed Sultan, Mirza Haider Jivan Beg, Akhtar Sultan, Akbar Beg, Mirza Abbas Beg, and Mirza Yahya Beg.

19. Begum Sarbuland reportedly spoke Urdu with a Delhi accent rather than a Hyderabadi one. Interview with Najeeb Jung, March 6, 2023. In keeping with her lineage connecting her to the Mughal court, she was punctilious regarding Urdu grammar. Her grandson Abid Ilahi recalls that even late in life, when she was afflicted with dementia, she might still rebuke him if he spoke the language incorrectly. Interview with Abid Ilahi, September 28, 2023.

20. Leonard, "Mulki-Non-Mulki Conflict."

21. These intrigues are described in a lengthy autobiography. There are two editions of the book. The first published version appeared in English, translated by his son Nawab Jivan Yar Jung. The second version, written in Urdu and edited by another son, Zulqadar Jung, contains significant differences. Agha Mirza Beg, *Karnama-e Sarvari*; Server-ul-Mulk, *My Life*. For a detailed study of a related controversy, see Cohen, *Appeal*.

22. Cohen, 256.

23. Minault, *Secluded Scholars*, 24.

24. Quoted in Naim, *Texts and Contexts*, 206–7, 211.

25. By way of comparison, one of her contemporaries was the mother of Munshi Zakaullah, a renowned scholar and the dear family friend of Begum Sarbuland's father-in-law, Samiullah Khan. Their family had tutored the Mughal family for generations. She, however, was entirely illiterate. Minault, *Secluded Scholars*, 25.

26. Minault, 205.

27. Akhtar, *Bibi's Room*, 56.

28. Minault, *Secluded Scholars*, 105–57; Akhtar, *Jara'id-e Nisvan*.

29. For a critical review of the history of Hyderabadi women's writing in Urdu, see Tahsin, *Haidarabad Meñ Urdu*.

30. Akhtar, *Bibi's Room*, 60–61.

31. Her private travel diary might be compared to that of another Hyderabadi woman, Muhammadi Begum (1911–90). The latter's journal includes significantly more verses in both Urdu and Persian. See Muhammadi Begum and Husein, *Long Way*.

32. Her granddaughter Shakira Sultan relates one of her devotional rubais in praise of Medina: "Zahe Nasib Madina maqam ho jae / Dar-e habib peh mera salam ho jae / Madina ja'uñ, dobara phir ja'uñ / Umar yuñ hi tamam ho ja'e."

33. For an introduction to the school and its founding and functioning, see Minault, *Secluded Scholars*, 206; Pernau, "Schools for Muslim Girls."

34. Akhtar, *Bibi's Room*, 58.

35. In fact, they were cousins, though several times removed. Begum Sarbuland's paternal grandmother was the cousin of Hamidullah's father, Samiullah, on her father's side.

36. Maulvi Muhammad Zaka'ullah, *Savanih 'Umari*, 25. Ali Ahmad Khan was also a first cousin to Sir Sayyid Ahmad Khan. Hamidullah's father, Samiullah Khan, actually married two of Ali Ahmad's daughters. The first wedding took place several years before 1857, and the couple gave birth to three children. However, the uprising led them to flee their homes and take refuge in the shrine of Nizamuddin, where his wife and their three children fell ill and died. A few years later, he married his deceased wife's sister on November 4, 1861. The pair had two sons, Hamidullah and his brother, Majidullah. She died in September 1894 and was buried near the shrine of Shah Abd ul-Aziz in Delhi. Maulvi Muhammad Zaka'ullah, 26–27, 183–85. In the four-hundred-page book dedicated to the life of Samiullah, his dear friend Zaka'ullah does not mention his wives at all, except to share their father's name and the dates of their deaths.

37. Unlike his wife, Samiullah traced his family lineage back to Arabia and ultimately to Ali ibn Abi Talib, the fourth Muslim caliph. *In Memoriam*, 10. Nevertheless, his ties to the Mughal family are indicated by his mother's lineage.

38. *In Memoriam*, 24. His own father had played a similarly critical role as the mir munshi for General David Ochterlony and Captain George Birch. Their faith in his work surely contributed to the government's later enthusiasm for entrusting Samiullah Khan with sensitive imperial tasks. Hamidullah was in turn given similar roles because he came from this lineage. Maulvi Muhammad Zaka'ullah, *Savanih 'Umari*, 8–10.

39. H. Khan, *Early Life*, 2.

40. H. Khan, 3.

41. For a description of this tour, see Iqbal Ali, *Safarnama-e Panjab*.

42. Lelyveld, *First Generation*, 167.

43. Lelyveld, 168.

44. H. Khan, *Early Life*, 8.

45. He was escorted there by his father, Samiullah, who kept a detailed journal of the trip in Urdu, which was later published in a magazine in Aligarh, then as a standalone book, and later translated into English and Tamil. See S. Khan, *Safarnama-e Maulvi*; S. Khan, *Musafiran*. His son also kept a written record of the journey in English, portions of which were later published. H. Khan, *Early Life*, 10.

46. Many of his travel diaries from Cambridge were later published. See the bibliography for a partial listing.

47. For an account of this trip, see H. Khan, *Leaves*, 31–42.

48. Excerpts of his travel diary were published as a standalone book, as were further extracts at a later date. H. Khan, *Early Life*, 28–28.

49. M. Khan, *Life and Times*.

50. Maulvi Muhammad Zaka'ullah, *Savanih 'Umari*, 175.

51. Danapuri, *Sair-e Dihli*, 14.

52. Maulvi Muhammad Zaka'ullah, *Savanih 'Umari*, 181–82.

53. Maulvi Muhammad Zaka'ullah, 182.

54. Storycore Archive, "Syed Family Interview."

55. Also spelled Mahomed Ullah and Mahommed Ullah. He, like his father, graduated from Cambridge and became a lawyer in Allahabad. He wrote several popular works on Muslim family law but died young, probably of cancer. He married a maternal cousin, the daughter of Zulqadar Jung, in 1916. *Tampa Morning Tribune*, "Elephants and Motorcars." He was alive as of June 1937. "India House Reception," 5.

56. Khurshid Begum, or Khurshid Iqbal Shahrbano, was the couple's eldest daughter and a lifelong companion for her mother. She attended Mahbubiya Girls' School and was educated to a "junior Cambridge" level. Her husband was Hamidullah's maternal cousin Abdul Majeed Khwaja (1885–1962), who is mentioned frequently in the diary (using an earlier form of his name, Khwaja Abd ul-Majid). They had nine children, three boys and six girls. She was one of the first women in her family and in Aligarh to leave purdah and was active in the anticolonial movement. Her son recalled that she "was torn between divided loyalties to her father [Hamidullah], a great Westernized liberal aristocrat, and her husband who, under Gandhiji's inspiration, had made a bonfire of his expensive and fashionable English suits and completely switched over to *khadi* kurta and pajama. She founded and ran successfully a Women's *khadi bhandar* (homespun store) at Aligarh and also edited an Urdu magazine, *Hind*. Gandhiji wrote his first ever letter in Urdu to Mother on a postcard." Khwaja, *Vision*, 5. She also founded a school, Hamidia Girls' School, for poor Muslim girls in Allahabad. She attended reformist conferences with her mother until late in the latter's life. She is buried in Aligarh.

57. He married the daughter of Begum Sarbuland's brother, Haidar Jivan Beg. Born in the mountain town of Solan, he became a civil engineer in Hyderabad.

58. He passed away in Karachi.

59. See note 98 below.

60. She married Mirza Kulich Beg, son of Begum Sarbuland's brother Haidar Jivan Beg. He died soon after his marriage and had one daughter, Shakira Sultana, who married Ajmal Khwaja, the son of Khurshid Begum and Abd ul-Majid. Tehzib Begum passed away in Medina and is buried in the Jannat al-Baqi cemetery.

61. Peile, *Biographical Register*, 2:691; "System of Administration." By rough comparison, in the city of Patna at this time, "a college student in the 1910s would live on about Rs. 20 monthly, a clerk Rs. 40, and a European engineer about Rs. 250." Boyk, "Provincial Urbanity," viii.

62. His retirement came a few months after the death of Nizam Mir Mahbub Ali Khan. While there is no firm evidence of this, it is possible that this change of rule may have contributed to the decision to step down.

63. A few years later he was succeeded by Begum Sarbuland's brother Haidar Jivan Beg, who was given the title Jivan Yar Jung. The later also studied at Cambridge and occupied the same role between 1918 and 1937. In this way, Begum Sarbuland held an immediate family relation with high-ranking Hyderabadi officials for at least half a century.

64. *Coronation Durbar-Delhi 1911*, 203.

65. Saberah Malik, personal communication, October 1, 2023. The blouse mentioned here has been preserved and is today held in a family archive in Lahore.

66. *Tampa Morning Tribune*, "Elephants and Motorcars."

67. His son-in-law Abdul Majeed Khwaja wrote a detailed biography of his life that was published in 1916. The years afterward were not well documented. Surprisingly, given his stature, even obituaries are hard to come by.

68. Lambert-Hurley, "Contesting Seclusion," 133.

69. *A Short Summary*.

70. *A Short Summary*; Mirza, *Roznamchah*, 5, 10. I am grateful to Siobhan Lambert-Hurley for helping me to locate these references.

71. Lambert-Hurley, "Contesting Seclusion," 137.

72. Lambert-Hurley, 147.

73. Mirza, *Roznamchah*, 6.

74. Shervaniya, *Zad us-Sabil*, 7–8.

75. Their home in Delhi was Hamid Manzil, a house near Delhi Gate that Hamidullah bought in the 1880s. The house was demolished in 1973 to give way to the Golcha Cinema, itself now defunct but still easily locatable on the main road leading toward the Red Fort. A portion of the original land still remains in the family.

76. Interview with Najeeb Jung, May 23, 2022.

77. Begum Sarbuland Jang, *Dunya*, v. See also pg. 5.

78. H. Khan, *Early Life*, 44. By a similar token, Begum Sarbuland seemed to view animals with compassion. A family story, related by Saberah Malik, relates that Begum Sarbuland once saw her granddaughter Taj Begum shoot two passing birds. "Begum Sarbuland Jung immediately chastised her by saying, 'what had those helpless birds done to you for you to take their flight and life?' My mother [Taj Begum] put down her guns and never fired another shot again. I remember joining B. S. J. [Begum Sarbuland Jung] on her terrace to feed birds, mostly urban pigeons." Personal communication, Saberah Malik, October 1, 2023.

79. H. Khan, *Leaves*, 31.

80. H. Khan, *Early Life*, 25, 44–45.

81. "Social and Personal (1925)"; "Social and Personal (1926)"; "Social and Personal (1932)."

82. *Indian Listener*, April 7, 1936. Entry for April 20.

83. I have deduced this date by comparing the book's preface to historical sources. In the preface, she notes that her husband had passed away. This indicates a publication date after 1930. She also writes that she had been to Medina twice, but historical sources confirm that she went to Medina at least three times, in 1910, 1934, and 1937 (see below). Thus, she certainly published the diary after 1930 but before 1937 and almost certainly after 1934. Thus, the likeliest dates are 1935–36.

84. Interview with Shakira Sultana, May 27, 2022.

85. Interview with Shakira Sultana, May 27, 2022.

86. While in detention as a prisoner of war, he reportedly treated a Japanese general, Yamashita. As a reward, he was allowed to order the latest American medical journals and writing paper. "Because of wartime shortages Ilahi Bakhsh was instead given an unlimited supply of toilet paper rolls. The doctor returned to his family in Delhi three and a half years later with a suitcase filled with his manuscript on toilet rolls." After 1947, Dr. Ilahi was later appointed as the personal physician to Muhammad Ali Jinnah, the founder of Pakistan. Abid Ilahi, personal communication, November 4, 2023.

87. Interview with Muazzam Ilahi, October 26, 2023.

88. Interview with Shakira Sultana, May 27, 2022.

89. Interview with Abid Ilahi, September 28, 2023.

90. Boyk, "Fatima Begum," 160. On Nishat un-Nisa Begum Hasrat Mohani's pilgrimages, see Majchrowicz, "Begum Hasrat Mohani."

91. Fatima Begum, *Hajj-e Baitullah*, 33. Though Fatima Begum does not mention it, Begum Sarbuland was accompanied by her granddaughter Taj Begum, daughter of Khurshid Begum. Personal communication, Saberah Malik, October 1, 2023.

92. The dates of travel suggest that, if the group went to Mecca after Medina, they would have performed an umrah (or "minor pilgrimage") rather than a hajj.

93. Lady Abdul Qadir (1884–1965) is best known today for her leadership roles in the Pakistan movement in colonial India. Her husband was Sir Abdul Qadir, a prominent judge and founder of the Urdu literary magazine *Makhzan*.

94. This refers to Halimullah and Khurshid Begum's daughter, Akhtar.

95. Miss Clements was Begum Sarbuland's private secretary. That Begum Sarbuland kept a British employee was a mark of status. That she was a convert to Islam would have been a further mark of pride. Interview with Najeeb Jung, May 23, 2022.

96. "Jedda Report for July 1937," doc. 183, E4918/580/25, §288; Bidwell, *Foreign Affairs*, 12:282. The final line refers to an investigation by local religious authorities as to the sincerity of Miss Clements's conversion.

97. He was visited around the same time by another prominent Hyderabadi reformist, Begum Sughra Humayun Mirza, who published an account of their meeting in her travelogue of Europe. See Mirza, *Safarnama-e Yurap*, 2:152–53; Lambert-Hurley, Majchrowicz, and Sharma, *Three Centuries*, 451–60.

98. In 1938, Halimullah was married to a cousin, Razia Sultana (1922–2016), a poet. Their children include Omar Farooq Jung (b. 1937), Akbar Jung (b. 1942), Zafar Hameed Jung (d. 2022), and Najeeb Jung (b. 1951). Halimullah left the medical track to become a businessman. He was CEO of Deccan Airlines and owner of a furniture store (Brooke's), several drug stores (Redicura), and two popular restaurants in Delhi's Connaught Place. These were Davico's and Piccadilly, the latter run by his wife, who also sometimes served as bartender!

99. Homes belonging to Muslims who were suspected of having fled to Pakistan were confiscated and allotted to Hindu and Sikh arrivals from Pakistan. Zamindar, *Long Partition*.

100. Interview with Muazzam Ilahi, October 26, 2023; and Abid Ilahi, October 27, 2023.

101. Interview with Najeeb Jung, May 23, 2022; Abid Ilahi, with Farooq Jung, personal communication, October 6, 2023.

102. Halimullah and his wife were later buried in an adjacent plot.

103. Storycore Archive, "Syed Family Interview."

104. Interview with Shakira Sultana, May 22, 2022.

105. Interview with Shakira Sultana, May 22, 2022.

APPENDIX 1

TRANSLATION OF AN ADVERTISEMENT FOR *THE WORLD THROUGH A WOMAN'S EYES*

Initiative for the Instruction of Young Women[1]
Begum sahiba will spend all proceeds from the sale of this travelogue
On the beneficial effort that she has prepared for
The Instruction of Young Women

* * *

GIRLS TODAY ARE BEING WASHED away by the relentless flood of the West, which is destroying their Eastern particularities, the true cause of which is that they are being given the wrong form of education. This incorrect form of education has not only placed their lives in danger but has darkened India's future as well. Sensing the destruction of the country and its daughters, Begum sahiba Nawab Sarbuland Jung has designed a tremendously beneficial initiative for the education and training of young girls. The initiative aims to open training centers for girls both in Delhi and outside of Delhi that will be able to give the girls not just a higher education but also teach them skills that will free them from ever having to worry about their own subsistence. These learning centers will also imbue the girls with proper morals and human kindness. This important initiative cannot be successfully achieved until the people of the country prove themselves willing to cooperate with Begum sahiba in the realization [of this project]. To provide financial support for these training centers, Begum sahiba has committed all the income from the sale of this travelogue for the benefit of these women's schools. All those who wish to support girls' education should offer their encouragement to Begum sahiba by assisting in the distribution of this travelogue, and all those who are interested in the important matter of girls' education should send a card to this address to request [information] about this initiative to provide training and instruction to girls: Begum sahiba Nawab Sarbuland Jung, Daryaganj, Delhi. —The Publisher.

Note

1. This text appears on the back page of the published travel diary (fig. 12.1).

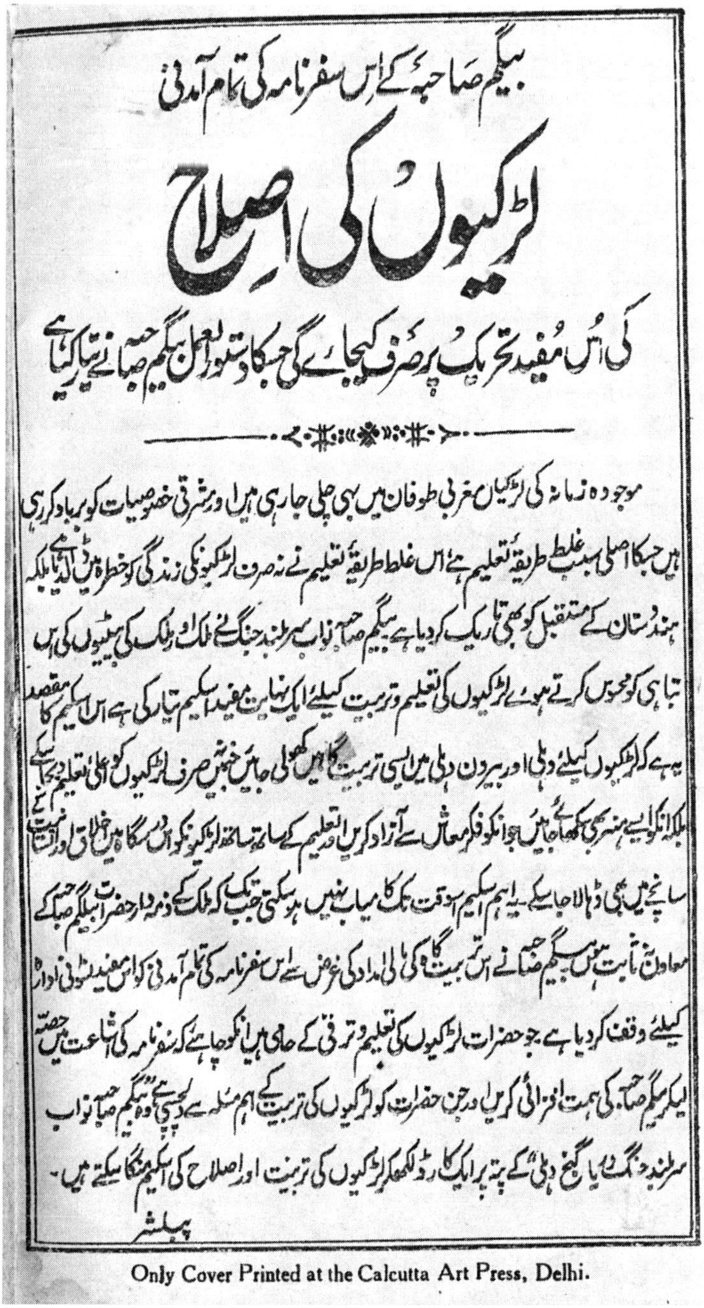

بیگم صاحبہ کے اس سفرنامہ کی تمام آمدنی

لڑکیوں کی اصلاح

کی اس مفید تحریک پر صرف کیجائے گی جسکا دستور العمل بیگم صاحبہ نے تیار کیا ہے

موجودہ زمانہ کی لڑکیاں مغربی طوفان میں بہی چلی جارہی ہیں اور مشرقی خصوصیات کو برباد کر رہی ہیں جسکا اصلی سبب غلط طریقہ تعلیم ہے اس غلط طریقۂ تعلیم نے نہ صرف لڑکیوں کی زندگی کو خطرہ میں ڈالدیا ہے بلکہ ہندوستان کے مستقبل کو بھی تاریک کردیا ہے۔ بیگم صاحبہ نواب سربلند جنگ نے ملک اور ملت کی بیٹیوں کی اس تباہی کو محسوس کرتے ہوئے لڑکیوں کی تعلیم و تربیت کیلئے ایک نہایت مفید اسکیم تیار کی ہے اس اسکیم کا مقصد یہ ہے کہ لڑکیوں کیلئے دہلی اور بیرون دہلی میں ایسی تربیت گاہیں کھولی جائیں جنمیں صرف لڑکیوں کو اعلیٰ تعلیم دیجا سکے بلکہ انکو ایسے ہنر بھی سکھائے جائیں جو انکو فکر معاش سے آزاد کریں اور تعلیم کے ساتھ ساتھ لڑکیوں کو اس درسگاہ میں اخلاق اور انسانیت کے سانچے میں بھی ڈھالا جاسکے۔ یہ اہم اسکیم اسوقت تک کامیاب نہیں ہوسکتی جبتک کہ ملک کے ذمہ دار حضرات بیگم صاحبہ کے معاون ثابت نہوں۔ بیگم صاحبہ نے اس تربیت گاہ کی مالی امداد کی غرض سے اس سفرنامہ کی تمام آمدنی کو اس مفید تعلیمی ادارہ کیلئے وقف کردیا ہے۔ جو حضرات لڑکیوں کی تعلیم و ترقی کے حامی ہیں انکو چاہئے کہ سفرنامہ کی اشاعت میں حصہ لیکر بیگم صاحبہ کی ہمت افزائی کریں اور جن حضرات کو لڑکیوں کی تربیت کے اہم مسئلہ سے دلچسپی ہے وہ بیگم صاحبہ نواب سربلند جنگ دریا گنج دہلی کے پتہ پر ایک کارڈ لکھکر لڑکیوں کی تربیت اور اصلاح کی اسکیم منگا سکتے ہیں۔

پبلشر

Only Cover Printed at the Calcutta Art Press, Delhi.

Fig. 12.1 A publisher's announcement on the back page of *The World through a Woman's Eyes* that commits proceeds from the book to the construction of a girls' school being established by Begum Sarbuland Jung. *Source: Daniel Majchrowicz.*

APPENDIX 2

SELECTIONS FROM THE TRAVEL ACCOUNTS OF HAMIDULLAH NAWAB SARBULAND JUNG

Nawab Sarbuland Jung published several travel accounts based on his 1909–10 journey to the Middle East and Europe with his wife, including three books and several newspaper articles. These were written variously in English and Urdu between 1911 and 1914. These works in English and Urdu are not translations of each other. Rather, each work focused on a different moment in the journey and was carefully written to cater to distinct audiences. Thus, in his English-language travel accounts, which seem intended for a British readership, Nawab Sarbuland styles himself as an anglicized intellectual in the mold of Richard Burton who could, in his own words, "assume the incognito of ordinary Indian pilgrims" and maneuver the "old prejudices" of the Muslim world. By contrast, when writing in Urdu for a presumed Muslim readership, he feels no need to "assume" such a guise but rather embodies it. In Urdu he enthusiastically evinces the very same "old prejudices" and preferential treatment for Muslims that he so wryly invoked in English, as when he dedicates his *Safarnama-e Qustuntuniya* to the Ottoman "martyrs" of the Balkan wars, "for whom the world's Muslims feel great sorrow due to their shared Islamic brotherhood" (fig. 13.1). Each of his accounts thus reveals new details about the couple's trip and his responses to travel in the Middle East and the Ottoman Empire as a man who consciously straddled Muslim, Indian, and European intellectual worlds and who shifted his self-presentation according to his audience.

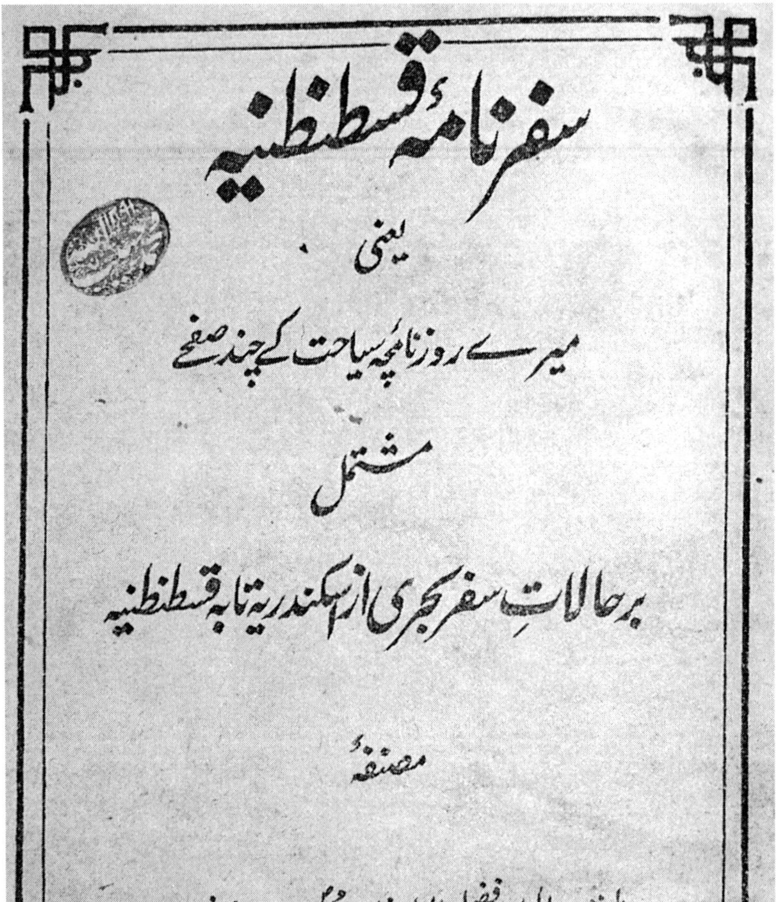
سفرنامہ قسطنطنیہ

یعنی

میرے روزنامچہ سیاحت کے چند صفحے

مشتمل

بر حالات سفر بحری از اسکندریہ تا بہ قسطنطنیہ

مصنفہ

عالیجناب الحاج افضل العلما نواب محمد حمید اللہ خانصاحب

سربلند جنگ بہادر ایم۔ اے کیمبرج یونیورسٹی و بیرسٹرایٹ لا

لنکنزان

قاسم پریس حیدرآباد دکن میں طبع ہوا

Fig. 13.1 The cover page of Hamidullah Khan's travel account, *Safarnama-e Qustuntuniya. Source: Daniel Majchrowicz.*

Nawab Sarbuland's collected works on the trip run to well over one hundred pages and cannot be included here in their entirety. This appendix brings together a selection of his English language writings published between 1912 and 1914. While these pieces are far less frank and unguarded than his wife's account, they nevertheless offer a fascinating counterpoint to her writing and a further window into the nature of their relationship. They draw particular attention to these authors' different focuses and interests. There are very few overlapping moments between their accounts. One significant difference between them is that, while she mentions her husband and her servant repeatedly, he almost completely excludes both women from his writing. In the passages below, his wife is mentioned directly only once, and their servant Amina bi, not at all.

All the selections have been lightly edited for grammar and consistency with the style used to render Begum Sarbuland's account above and to modernize the spellings where necessary. These selections follow the couple's travels through Palestine, the Sinai, and Alexandria.

Excerpts from *A Pilgrimage to the Near East*

A Pilgrimage to the Near East *was published in Secunderabad in 1912. Written in English, it is his longest work on this journey. The account focuses on his time in and around Suez. It is styled as a reflection on the changes that had occurred since his time living in the region in the 1880s. The following are selected excerpts from the book.*

Arrival in Suez

Five and twenty years have rolled by, the young attaché has become a man of the world,[1] and embarks from Bombay in the P&O SS *Salsette* at 2:30 p.m. on November 13, 1909. Aden is reached on November 17. From the distance, this place has been seen by all European travelers and needs no introduction. The passengers here transhipped to the SS *China*, which left about 7:15 p.m. the same day and approached Suez on November 21.

The land of the Nile is once more within sight. It is a beautiful day, though somewhat warm. What changes have taken place in Egypt since 1884? HH the Khedive Taufik Pacha has been succeeded by his son HH Abbas Pacha Hilmi; the noble (Thomas George) Earl of Northbrook GOSI is no more;[2] Earl Cromer (formerly Sir E. Baring) PC, GCB, GCMG, etc., after a long and successful career has retired from Egypt;[3] Nubar Pacha's place is occupied by Boutros Pacha Ghaly. Minor lights, who played important parts in the making of Egypt, have also done their work and gone.[4]

As we intended to return to Suez, we left some of our heavy luggage with the agent of the Khedivial Line of Steamers, who had come to the *China*. For the sake of convenience, Egypt may be regarded as our center, whence short excursions were made between November 1909 and March 1910.[5]

The sea remains the same old "Reedy Sea"; but the approaches to the Suez Canal have been greatly improved. There is some talk about increasing the width of the Suez Canal, and excavation works are going on. In 1884 there was no traveling by night through the Canal, now we take electric light on board the steamer and go on without stoppages.

[NOTE:—We have traveled by water from Suez to Port Said and vice versa, and have had the advantage of practically seeing the entire 160 kilometers of the Canal by daylight. For the sake of convenience, we have put together the observations of both the journeys in one place.]

We passed the Gulf of Suez, which is extremely shallow, and proceeded through the Canal towards Port Said. The monument of Darius has been located about the 139th kilometer. Some time ago near Shaluf (139th kilometer), which comes next, were discovered crocodiles' teeth and the remains of hippopotami in the sand. Also, in the next layer of the stone, vertebrae of sharks, bivalve shells, and the remains of the bryozoa. We saw two iron lighthouses 65 feet high. Then passed the two Bitter Lakes, which Brugsch has identified with the Marah of the Bible. We entered Lake Timsah (area 6 ½ miles) and passed the Gebel Maryam. We got a view of the mountains of the Peninsula of Sinai and traversed Lake Bolah. Close by, Professor Petrie carried out some excavations in 1886 and identified the site of Tehaphnehes of the Bible. We reached El-Kantara, which used to be on the ancient caravan route from Egypt to Syria. We passed through the Lake Menzaleh and reached Port Said.

We landed at Port Said and assumed the incognito of ordinary Indian pilgrims. Burton has described with great humor his adventures in obtaining a passport. The Indian pilgrim, who had previously obtained a passport in India, but who out of foresight obtained another at Port Said, had very little difficulty in obtaining it, the whole process taking a couple of hours.

Port Said in 1884 had a population of about 18,000, but now boasts of over 50,000 inhabitants, about a fourth being Europeans. The harbor occupies 570 acres and is protected by two piers. The lighthouse (174 feet) constructed of concrete being one of the largest in the world. The colossal statue of Ferdinand de Lesseps, by the sculptor Fremiet (unveiled in 1899), is an object of admiration for all travelers.

The stately dome of the palatial Suez Canal Company's offices, the beautiful quay, the neat-looking streets, the solid blocks of commercial buildings have replaced the poor surroundings of the Arab town of the olden days. Rinks, theaters, hotels, electric tramways, and other accompaniments of the modern civilization have much improved the town. Extensive shipping works and railway stations have sprung up. A railway runs alongside the Canal from Port Said to Port Taufik, with branches to Cairo and other places. Everything is up-to-date, including dining cars and motor vehicles. Port Said has entirely lost its old character and except for the existence of an Arab quarter, and the sight of Egyptian faces, one might almost imagine oneself in a European port of the Mediterranean. Though outside appearances are European; there is still Egypt behind the veil.

East is East and West is West,
Never the twain shall meet [*sic*].

Remembering the old prejudices, we put on a fez when landing from the *China*. The effect is magical. We gained the sympathy of the children of the land by talking to them in Arabic. As soon as they found out that we were Muslims they opened their hearts to us. Verily, the bond of religion is strong. They talk far more English now than they did in 1884, but French has not yet lost its hold upon the people. Even in "English" places you find the clerks preferring the use of French.

Things progress more smoothly if you can talk Arabic to the Egyptians, and French, English, or German to Europeans, according to their nationality. "God be praised," "I am a Muslim," at once secures for you the land. Where a Christian would be in danger at Port Said, a Muslim is safe. Merchants lower their prices, your credit goes up with a bound, your word is trusted, and the respect shown to you becomes genuine. In any case of difficulty, you find yourself as if you were amongst sympathetic friends. Every Muslim house is opened to you and every Muslim ready to assist you.[6]

A Description of Suez

The food of the Suezians is not so characteristically Oriental as it was in Burton's time, now foreign products are largely consumed. . . . Burton has described Suez, at the time of his several visits. Those who have read Chapter IX of his amusing work will hardly thank us for going into details of life at Suez; but his account must be accepted with a grain of salt.[7] According to him the population was between 1,800 and 6,000; but Suez has now a population of over twenty thousand, including some 3,000 Europeans, about ten mosques, large docks, and extensive warehouses. The town has a governor and is now kept clean. To the north of the town the khedive has built a kiosk which commands a fine view. Close by is the mouth of the freshwater canal, and not far off the waterworks. According to Greek and Roman historians there existed a navigation canal 62 miles long in these parts; but the present canal, which in most parts is a restoration of the old one, is mainly used for the supply of drinking water.

Traveling through Sinai and Egypt by Train

The site of Pithom of the Bible has been located near the neighboring ruins of Tel-el-Maskhuta. Kasasin and Tel-el-Kebir, noted as the scenes of British victories in 1882, are also on the line. When passing this way in 1884 we remembered seeing the remains of the field of battle; bullets could be picked up in those days and shells imbedded in walls and roofs were plentiful; some of the fortification works had not till then been quite leveled.

Most of the railway stations in the olden days were without raised platforms, but now they equaled English stations in style and finish; peace reigned supreme everywhere, and the old places had such a prosperous look that one could have hardly identified them with the battlefields of yore. But the sky and the desert were the same and the dust in the railway carriages as bad as described by Sir Donald Mackenzie Wallace in his excellent work on Egypt. At Abu-Hammad begins the Eastern Delta, which is even more fertile than the Western Delta. Zagazig is soon reached.

In Alexandria

We have paid many visits to the City of Alexander and found the place ever changing. We do not propose to describe at length its sights or to go into details of a political nature but will confine ourselves to a brief survey.

The Jews have been here since the fourth century BC and as long ago as 48 years before Christ, Alexandria had more than half a million inhabitants, the Greeks predominating.

In October AD 641, Amr-ibn-e-'As, the great general of the famous Khalif Omar, conquered the city. Although he treated the inhabitants with great moderation the non-Muslims of Alexandria have ever since passively resisted Islamic rule. The Greeks have never forgotten their olden days. In AD 1800 only 4,000 inhabitants had been left on Alexandria; but the town improved again, and by 1880 when we first went there, it had assumed large dimensions. Burton calls it "a city of misnomers, whose dry docks are ever wet and whose marble fountains are eternally dry, whose Cleopatra's

Needle is neither a needle nor Cleopatra's, whose Pompey's pillar never had any earthly connection with Pompey."

In 1880 we had found the site of the Serapeum, and the so-called Pompey's pillar (which according to the latest theory was put up by Emperor Theodosius in AD 391) surrounded by filth and rubbish. The Indian Mails in those days went by railway from Suez to Alexandria and thence by steamer to Brindisi. Alexandria was quite a large place but not so much Europeanized as now. Some of the streets were well kept and Mehemet Ali Pacha's Palace, which we saw by special permission, formed its greatest attraction. The Arab quarter was avoided by peace-loving tourists.

The bombardment of 1882 reduced the growing town to a most pitiable state, and tears came into our eyes when we saw it afterward. Even up to 1884 Alexandria had recovered from its misfortunes. Its best quarters were in a state of transformation and the town looked a forest of scaffolding. Thanks to mercantile enterprise the Place Mehemet Ali has become once more a beautiful square and over a dozen well-paved roads vie in elegance and wealth with the streets of European towns.

To the pilgrims, however, the shrine of Daniel, the Prophet, is pointed out at Alexandria, as formerly, though the Persians place the same tomb at Sus; and the tombs of Mahomed El-Busiri; the author of the famous poem called Al-Burdah, and of Abu Abbass El-Andalusi, are also shown.

In 1900 Schiess Pacha and Dr. Botti excavated the catacomb on the Kom-Esh-Shukafa. The Greek Rock Tombs, dating from 200 BC, to the North of the Ras-et-Tin palace have been lately (since 1901) cleared out. The splendid Museum of Graeco-Roman Antiquities, founded in 1903, has been enriched by the finds of excavators and collectors like Botti, Sieglin, Breccia, and others.

Music in Egypt

In every land, music hath its charms, and Egypt is no exception. The usual instruments are the tambourine with little bells, the semi-spherical tambourine, the haut-bois, the drum, the kettledrum, and the funnel-shaped drum made of earthenware or of wood, inlaid with mother-of-pearl and tortoise shell, with a fish skin stretched over the broad end. A better class of instruments used for chamber music includes the flute, the two-stringed violin, the one-stringed violin, the zither, and the mandolin.

The common basic element of all music is the harmony of sound, but there is a fundamental difference in the ideas of the East and the West. What will appeal most as a concord of sweet sounds to one may repel the other. Their music and songs are the outcome of their national characteristics. The songs are generally of a lyrical religious description. The pleasures of friendship and enjoyment are sometimes made the theme of song, derision of an enemy, or contempt for the rustic fellah is expressed on other occasions.

Attitude toward Non-Muslims

On page 130, vol. II, his Lordship (Cromer) says: "On the other hand, the annals of Islam are not stained by the History of an Inquisition. More than this, when he is not moved by any circumstances especially calculated to rouse his religious passions, the Muslim, readily extends a half-contemptuous tolerance to the Jews and Christians. In the villages of Upper Egypt, the crescent and the cross, the mosque and the monastery, have stood peacefully side by side for many a long year."

This is the key to the attitude which Muslims have assumed against non-Muslims. Islam, as a worldly power, rose at the expense of Judaism and Christianity and, in spite of its toleration, the conquered people never forgave the followers of the Prophet for having substituted Muslim governments for their own. Islamic rule was a tolerant rule, because every true Muslim believed that it was the will of God that, up to a certain time, people of different religions should exist. No Muslim could, therefore, refuse to tolerate those who professed other religions. What is assumed to be non-toleration is really the exclusiveness of Islamic doctrines and the fact of their not admitting of any compromise with Pantheism. As a political force Islam has always been tolerant. True Christianity is the same as Islam. Love and charity are as great virtues in Islam as in Christianity; but Islam puts the will of God and the command of God higher than everything. Attributing all improvements in Europe to Christianity and all backwardness in the East to Islam is to ignore the characteristics of race, climate, and country. The comparison of the character of Muslims and non-Muslims, as a whole, in the same country and of the same standing will be to the advantage of the former. The real cause of the Englishmen's difficulty is the jealousy of powers and the self-interest of the Egyptian factions.

Excerpts from Newspaper Publication of "A Pilgrimage to Medina"

Nawab Sarbuland published at least four articles in the Allahabad newspaper The Leader *in late 1913 and early 1914. Though the series is entitled "A Pilgrimage to Medina," in these extant articles he never reaches any further than Jordan. Instead, the series seems to continue in an English-language addendum to an Urdu travelogue,* Safarnama-e Madina.

Chapter I[8]

On a Monday in November 1909, at 6:00 p.m. we left Port Said by the Austrian Lloyd SS *Imperatrice*, which was not as tidy as the P&O boats. We had, however, a four-berth cabin all for ourselves. Here we met a retired Anglo-Indian officer who was now employed by Port Said. One of his duties being to inspect the Aust. Lloyd's vessels. A number of peddlers and fruit sellers, etc., boarded the steamer at Port Said, and some remained on the steamer. A large number of Syrians, Turks, Arbs, etc. were among the deck passengers. The evening meal was served at seven in the usual German style.

We conversed freely with our saloon fellow passengers, who talked Arabic, French or English, and found most of them imbued with the spirit of making money. The deck passengers had made themselves comfortable in all sorts of warm nooks and corners. A Turkish dame traveling with her brother as a deck passenger was scarred on the face, and on being questioned related how she had been once attacked by robbers, who had branded her in this manner and relieved her of all her belongings. She was now a sewing mistress employed in some Syrian school.

There were passengers both male and female, some Church functionaries, and others following commercial pursuits. Some of the ship's officers spoke French, and we were able to obtain some information from them about the new railway to Medina.

Jaffa

We reached Jaffa (the beautiful) before sunrise and anchored there all day, embarking passengers and discharging and loading cargo. Sheikh Ibrahim Hasan al-Ansari (of Jerusalem) boarded our steamer and offered us his assistance, but as we had no time to go to Jerusalem we declined his offer with thanks.

The surrounding views were most charming; the sunset magnificent. We sum up here the information obtained during our two visits to Jaffa. The Prophet Jonah is said to have just quitted this port when he was swallowed by the whale (Jonah 1–3). In the days of Solomon, it was the port for Jerusalem (2 Chronicles 11:16); after the end of the seventeenth century, a quay was constructed here. There is a very busy Arab bazaar here, various mosques and khans, the Latin Hospice founded in 1654, a public garden and a clock tower. Jaffa contains about 45,200 inhabitants and about 20,000 pilgrims pass through it yearly. The most conspicuous object is the tower of the Greek church rising above the flat-roofed houses of Jaffa, and the British Hospital built owing to the efforts of Miss Maugau and supported by voluntary contributions.

The steamer, which has to lie at some distance from the shore, is surrounded by a number of passenger boats and lighters for cargo. The scene is an animated one. We were struck by the skill displayed by the Jaffa boatmen in handling their craft. Passengers should look sharply after their luggage; they should avail themselves of the boatman belonging to some recognized tourist agency and make a bargain before leaving the steamer. The usual charge is two francs per person.

Jaffa provides for tourists, as well as for pilgrims on their way to Jerusalem, so every provision is made for travelers, during the pilgrimage and spring months. The house of Simon the Tanner is in a mosque near the lighthouse on the southern side of the town, where we recited some prayers. Orange gardens surround the town and the fruit forms the principal article of export. The trees are in full flower during the months of February and March, and the fruit ripens from the month of September onwards. Visitors should ask permission before entering the gardens.

There are German and Jewish agricultural colonies at Jaffa, Sarona, and Richon le Zion whose efforts are mainly directed to vine culture and whole grapes. The railway from Jaffa to Jerusalem has been completed in recent times and has been of great utility. There is only one train daily in each direction. In 1799 Napoleon I conquered Jaffa and laid siege to Acre. He defeated the Turks on the plain of Jezreel and penetrated as far as Safad and Nazareth. In 1840 Napier bombarded and captured Acre.

On the upward voyage, there is almost a straight line of coast to Haifa with the ancient ports of Apollonia and Caesarea, once the capital of Roman Syria, now almost filled up with sand. After passing the port of Caesarea, Mount Carmel is in full view, and below it Athlit or Castellum Peregrinorum, one of the main strongholds of the crusaders.

Chapter II[9]

Haifa

We reached Haifa (ancient Sycaminium) on Wednesday before sunrise. We saw Mount Carmel, which next to Zion has the most sacred associations for religious students. The monastery on the top of the hill was erected on the cave where Elijah took shelter from Ahab.

Haifa is a pretty little town, situated at the foot of Mount Carmel and at the southern extremity of the bay of Acre, population 16,000, half being Christians. There is a considerable German colony, to whose industry much of the present prosperity of the place is due, and as a port it has considerably diminished the importance of Saint Jean d'Acre. There is a considerable export in grain, oil, sesame, etc., extensive vineyards as well as soap factories.

The Carmelite Monastery, on Mount Carmel, is reached by an easy carriage road or by a shorter footpath in thirty minutes. It was used by Napoleon as hospital in 1799 when he was besieging Saint Jean d'Acre and was destroyed by the Turks in 1821, but afterwards, rebuilt. Owing to the heavy dew, which falls every night, Mount Carmel remains evergreen, an unusual thing in Palestine. It was known as the Mount of God, and Elijah (I Kings XVII) performed his miracle here.

From Haifa one can visit by carriage or on horseback the famous town of Akka, Saint Jean d'Acre (ancient Ptolemais) and it has a long record of sieges, battles, and conquests. It was the crusaders' principal landing place, and part of the fortifications date from that period.

The visitors generally return along the shore, crossing the rivers Kishon and Belus. The bay of Acre yielded the shellfish which produced the famous Tyrian purple. From Haifa one can drive in six hours to Nazareth. We engaged a landau through a tourist's agent with a Moslem driver. They use three horses instead of a pair. We drove through the suburbs of Haifa, traversed the plain of Kishon, running parallel with the new railway line, and the brackish springs of Es Sadi.

At the first village, three miles distant, we met a sheikh, who was evidently a great man in this neighborhood. He gave us names of Mufti Abdul Salam and Rais-ul Ballad Said Effendi, with some instructions as to what we should do at Nazareth and Tiberias. Our driver was much interested in him, and as there were some tombs there, like good Moslems, we said the fatiha.

Round about this village, Balad-es-Sheikh, we passed through beautiful olive groves. A well, known as Bir-Miryam, is also on the route, from which some Syrian women were drawing and carrying away water in their pitchers. They were good natured creatures and as our coachman took some water from them, we gave to one of the damsels the equivalent of a three-penny piece, to her great delight.

We met a horseman going the same way as ourselves and as he looked like a sociable person, we entered into conversation with him and offered him tobacco and some sweets. We also passed some springs, where the driver watered his horses again. Many women, with oxen or donkeys, were also watering their animals there. We gave them some eatables.

Then we passed some extensive mulberry plantations, the village El-Yajur, the Wadi Es-Shomariyeh and the Tell-Omar. We got a good view of the country round. After passing less important places, we reached Al-Harithiyah, Harosheth of the gentiles (Judges IV:2).

There were many oak groves in the neighborhood of Wadi Zeida, Semuniyeh, Jebata Malul Mujeidil, and Yafa (Japhia of Joshua XIX 12), which is quite a big place, with various churches and schools. Josephus once fortified this place, and a tradition arose that this was the home of Zobeidee and her sons, James and John. From the neighborhood of the village Yafa there is a good view of the road to Nazareth. The Latin chapel of Maria De Tremore is passed before reaching Nazareth. There is a tradition that the Virgin was standing here when the Jews wanted to cast Jesus down headlong from the top of the hill (Luke IV:2, Ex. 29).

We reached Nazareth at 3:00 p.m. and [illegible] . . .

Chapter III[10]

Nazareth

El-Nasira is situated on the south slope of the Jebel as-Sheikh (1,600 feet). Cactus hedges, fig, and olive trees abound. The population consists of about 4,000 Muslims, 4,000 Orthodox Greeks, 1,000 united Greeks, 1,000 Latins, 200 Marionites and 250 Protestants. We saw all the churches and prayed at the chief mosque.

The Church of the Annunciation, situated within the Latin Monastery, was completed in AD 1800. It is twenty-three yards long, sixteen yards wide, with a nave and two aisles. The vaulting of the nave rests on four arches borne by four massive pillars. Among its many paintings are (1) an Annunciation and (2) a Mater Dolorosa, by Terallio (Spanish). The crypt is below the high altar. A handsome flight of marble steps descends to a vestibule called the Angel's Chapel; on the right is the Altar of St. Joachim, on the left, of the Angel Gabriel. Between the two altars is the entrance to the Chapel of the Annunciation, with the Latin inscription, "Hic verbum caro factum est" [*sic*] (here the Word was made flesh). The round upright column of Mary, a fragment of red granite column depending from the ceiling, and said to be miraculously supported above the spot where the Virgin received the angel's message.

On the rock here is said to have stood the house of the Virgin, now transported to Loretto near Ancona. Adjoining is a dark chamber called the Chapel of St. Joseph, which contains the altar of the flight into Egypt, with the inscription, "Hic erat subditus illis" (here he became subject to them), close by is the old cistern called the Kitchen of the Virgin, the mouth of which is said to be the chimney.

We also saw the workshop of Joseph and the synagogue in which Christ is said to have preached (Luke IV:16 et seq). We crossed the market and proceeded uphill to the table of Christ, on the west side of the town. The Latin Chapel here was erected in 1861. The table is a block of hard chalk (111 by 91 feet), on which Christ is said to have dined, with his disciples, both before and after the Resurrection.

After late breakfast on a Thursday, we left for Tiberias, visiting the Church of Gabriel, of the Orthodox Greeks. This was built about the end of the eighteenth century and is half underground. The spring is conducted past the altar on the left side. The Greek pilgrims use the sacred stream for bathing their eyes and heads.

Mary's Well (Ain Miryam) is the only spring which the town possesses. From the hill of Weli Siman you get one of the finest views of the whole of Palestine; and the first thought that strikes you is that this must have been a favorite resort of Christ, who must have had constantly spread before him the great library of Biblical story.

Looking across to the west, you make out the beautiful Bay of Acre; the ridge running out into the sea is Mount Carmel, crowned with its convent; southward are the mountains of Samaria; southeast, the hills round Jenin; eastward, the mountains of Gilead; and in between lies the magnificent plain of Esdraelon, covered with the silver line of the ancient river Kishon; northward the view culminates in glory as Hermon, like a great wall of white crystal, stands out against the blue sky, with the Galilee hills below it, and everywhere round that region, scenery varied and picturesque.

From Nazareth to Tiberias it takes about five hours to drive. After leaving Mary's Well we go uphill of Al Khavak, from which there is a fine retrospect and pleasing view of the quiet home at Nazareth. Descending the hill we reach Reineh and then Al-Meshhheb, reputed to be the birth place of the prophet Jonah. We next reach Kafr Kenna (Cana of the Bible), where Jesus performed his first miracle feast.

The Latin Chapel occupies the site of a Church of the Crusaders, as was recently discovered during its enlargement, which in its turn had succeeded a still more ancient church. A Hebrew mosaic inscription (third or fourth century), in front of the altar, names a certain Joseph as its founder; possibly Count Joseph of Tiberias, a converted Jew who was created a count by Constantine the Great and built several churches. Some still earlier remains seem to have belonged to a synagogue, traditionally said to occupy the spot where the water was made wine (John II). In the Greek church stone jars are shown, which are said to have been needed on the occasion of the miracle. Tradition also points out the house of Nathaniel (John I:45).

[Chapter IV][11]

From Kafr Kenna the route leads to the northeast through the broad and well cultivated Wadi Rimmaneh, a side valley of the plain of Sahel el-Battof. We pass the ruins and water basin of Birket Meskana and in twenty minutes more reach the foot of the hill, on which lies the village of Lubiyeh. In 1899 the French under Junot fought heroically against the superior force of the Turks, near Lubiyeh.

We now cross a low saddle, where a fine view is obtained of the trough-shaped plain of Sahel-el-Ahmar and of the mountains beyond Jordan. Here lies the traditional Mount of Beatitudes, where Jesus preached the Sermon on the Mount. It is declared by Dean Stanley to meet all requirements of the Joseph narrative.

The road past Lubiyeh becomes bad and in some places wild. The road is joined on the right by that from Mount Tabor. In about twenty-five minutes we reach the ruins of the Khan Lubiyeh. The Karn Hattin becomes visible to the North; we go along the base of this hill and after fifty minutes approach the edge of the plateau, whence we have our first view of the lake [the Sea of Galilee]. Near Hattin, Saladin inflicted a severe defeat on the crusaders from which they never recovered. Safed lies to the North, high up on the mountain, and Tiberias itself becomes visible in half an hour. It takes three-quarters of an hour more to descend to the town. Descending the slope by a winding road, we finally emerge at a point where the Sea of Galilee, a thousand feet or more below, bursts on our vision. Here we said our maghrib prayers.

The lake of Tiberias, also known as the Sea of Galilee is 137 to 157 feet deep, about thirteen miles long and nine miles broad. Its surface is 681 feet below that of the Mediterranean Sea. It contains many kinds of fish, among them being the Chromis Simonis, the male of which carries the eggs and the young about in its mouth, the Coracinus of Josephus, and the Barbur of the Arabs, which emits a sound.

As we approached Tiberias by the carriage road, from Nazareth, at sundown, we first observed the sarai, with its many domes, to the left and the mosque to the right. The extensive ruins of the castle lie to the north; near it is a mosque, with a few palms. There are about ten Jewish synagogues and the tombs of various famous Jews. The Hotel Tiberias, where we put up, is well situated on the high land, surrounded with ruins and modern Syrian houses. Cook and Company have practically been the cause of its prosperity. The manager was not present when we reached there and a young lady of the family offered us two rooms for 10 francs each, including board. Soon after we had settled down and had our dinner, the manager's wife put in an appearance and apologized for the young lady's mistake in offering us rooms at the usual rate. She argued that as we were tourists, we should pay at a higher rate than the people of the land. Another resident at the hotel, who was dining with us, told us that he himself had paid at a lower rate and mildly remonstrated with the landlady, but her feminine logic was unconquerable. In this journey as a hajji [pilgrim going on hajj] I had made up my mind to quarrel with nobody, and so I told the landlady to please herself. We saw the town, explored the bazars, admired the view, and studied the old ruins. She knew Arabic; but as she herself could not give us reliable information about the Medina railway, she kindly took me to a Jew's house later on in the evening. He was a contractor, and probably a moneylender; we did not find him at home. Next morning when we met, he told us that he too wanted to go to Samakh. I saved our Jewish acquaintance his expenses by offering him a place in our landau, but like a true Hebrew, our new acquaintance returned the compliment by taking us to an extortionate Greek hotelkeeper at Samakh.

Tiberias is the chief town of a kada [*qaza*] of the liwa of Acre, with 6,000 inhabitants, mostly Jews from Poland.[12] Here every place is historically associated with the life of Jesus Christ. The Lake of Gennesaret, the most holy sheet of water in the land. With the exception of Magdala, the home of Mary Magdalene and the plain of Gennesaret, great uncertainty exists about the true positions of the [unclear] which lined the [unclear] in the . . .

In going north from Tiberias we have, after Magdala and the plain of Gennesaret, Capernaum Bethasida, and Chorazin all on the west shore of the lake.

At Tiberias we received a telegram in answer to one sent by us from Haifa to the following effect: "Khedivial agency Port Said accept your changing." This referred to the change of dates for our departure from Suez to Jeddah. The delivery peon waited for bakhshish and the hotel people backed him up, but I did not corrupt good morals.

Close by are several hot-spring baths, with separate buildings for different classes of people at different rates. The principal spring has a temperature of 143°F, the water contains sulfur and chloride of magnesium. We drove to these baths close to the tomb of the famous Rabbi Meir and his two pupils. We then passed some historical ruins and a new Jewish colony. Beyond the baths is a synagogue of the Sephardim and a school of the Ash-Kemazim.

Crossing the River Jordan, on a Saturday at about 5:00 p.m., when the water was only about three feet deep at the crossing, close to the ford, known as Rab-et-Tumm, we reached the railway station.

We explored the neighborhood and revisited this extremity of the Sea of Galilee and the river. The Jordan (Arabic: Esh-Sheriba el-Kebir, "the large watering place") rises on Mt. Hermon 1,700 feet above the sea. It has two main collecting basins, the upper at the lake of Huleh, and the lower and large one at the lake of Tiberias. Its main course, from the lake of Tiberias to the Dead Sea, has a fall of 610 feet, and a length, owing to its numerous windings, of upwards of 185 miles; while the airline distance between the two lakes is little more than sixty miles. The deep valley of the river is called El-Ghor by Arabs. From time immemorial this has formed a natural boundary, as the paths descending to the river are all wild and rugged. Most of the northern part of the valley is fertile, while in the southern part barren tracts alternate with green oases. Many of the tributary streams, particularly those on the eastern side, are perennial.

The older channel takes half an hour to cross. The present and deeper channel averages only one hundred feet in width, but the river often overflows its banks after rain. The thicket which conceals the water from view harbors wild boars and birds and was formerly infested by lions (Jerem. XLIX:19). The water is of a tawny color from the clay, which it stirs up in its rapid course, and its temperature is high. It contains numerous fish. In ancient days, as at present, the Jordan seems to have been crossed almost exclusively at its few fords (1 Sam. XIII:7; 2 Sam. X:9,31). The most famous ford is that of Mahadet Hajleh; another ford, El-Henu, lies further to the south.

Samakh is 53½ miles from Haifa by railway. The stationmaster bore the famous name of Saucrat (Socrates); his wife, mother-in-law, and sister-in-law became friendly to my wife. They have rooms over the station offices.

Beyond Samakh the line enters the mountains of the country east of the Jordan and commences to ascend the valley of the Yarmuk. This river, which derives its Arab name, Sheriet el Menadireh, from the Bedouin tribe 'Arab el-Menadireh, was known to the Greeks as Hieromyces, a corruption of Yarmuk, the name given to it in the Talmud. It descends from the Jebel 'Ajlum to the south. Its volume is nearly as great as that of the Jordan. The deep valley through which it flows penetrates rocks of limestone; but, after the channel had been hollowed out, the valley must have been covered with a stream of volcanic rock, through which the river had to force a new passage.

After twice crossing the river the line reaches El Hammi, with the famous hot springs of Gadara or Amartha. The sanitary properties of these springs are highly extolled by Eusebius and many other ancient writers, and they are to this day visited by many persons in April. The principal springs are situated in a small open space on the right bank of the river. Around the great basin, which is partly artificial, are traces of a vaulted bathhouse. The water smells and tastes of sulphur, and, though clear in appearance, deposits on the stones a sediment which is used medicinally. The Bedouins regard the bathing places as neutral ground. The ancient Gadara, now named Mukeis, lies on the height to the South of the river, one hour distant from the springs.

The line continues to follow the Yarmuk valley, recrossing to its south side. The passage of the narrow gorges with their steep rocky sides and (in the rainy season) rushing torrents present a series of picturesque views. A number of similar deep wadis debouch from both sides into the Yarmuk valley. Just before reaching the mouth of the Wadi Ain Ghazal (left; south), the line crosses again to the north side of the valley. It then threads several tunnels. The Wadi Kleit then joins the valley from the left (south). Station of Wadi Kleit (66½ miles). The Yarmuk is now joined on the right (north) side by the Nahr er-Rukkad, which rises on the southern slopes of Mount Hermon. Though this is the chief river of the Jolan, it is quite dry in summer. At the confluence of the two rivers, we are still 157 feet below the level of the sea.

Tell-el-Makarim lies at the junction of three important streams, the Wadi-Esh-Shallaleh, coming from the southeast; the Wadi-ez-Zeidi, from the east; and the main source of the Yarmuk from the northeast. The upper part of the last is known as the Wadi el-Ehreir. The line leaves the Yarmuk valley and ascends in deep and steep curves to Zeizum on the north side of the Wadi-ez-Zeidi, here named Moyet Zeizun.

On a Saturday about 1:00 p.m. we reached Dera, where we changed trains. It was very cold. There is a hotel close to the station and many shops.

From the hotelkeeper we purchased a few articles of food; but with the exception of coffee, none of them were up to much. We bought several other things here, including a cardigan jacket and a sheepskin cloak to serve as a rug.[13] The prices were about 50 percent higher than one would pay elsewhere. Our porter at this station was a gay Lothario, willing to accompany us wherever we would go; but we left him at Dera with a good tip. Some of the railway engineers here talked French and we got a good deal of information from them. From this place we telegraphed to Medina to the manager of the rabat where we intended to put up.

Dera (76½ miles from Damascus and 100 miles from Haifa) is the seat of kaimmukam[14] with 4,000 inhabitants. It is the ancient Ebrawi and during the Christian period was the seat of a bishop. In the Wadi Ez-Zeidi lies a large reservoir (65½ x 50).

Pilgrimage to Medina

In 1914, Nawab Sarbuland published a book on his pilgrimage to Medina. The book is largely in Urdu, but it includes several pages of text in English, which are reproduced in full below. These passages seem to be a direct continuation of the series of newspaper articles given above.

Most European tourists to Syria have traveled from Damascus to Haifa via the Dera' Junction of the Hijaz Railway. Dera' is the Ebrewi of the ancients, and during the Christian period was the seat of a bishop. Finding ourselves there in November we proceeded to Medina as pilgrims. Trains run only thrice a week to Medina in the busy season. Ours left at 5:00 p.m. on a Saturday.

Crossing the Wadi-ez-Zeidi and skirting the side of the Zumleh we reached Kala Mefrak, where the railway line meets the old pilgrim route. Passing Fort Zerka and the upper end of the wadi of that name, we ascended the Wadi Aman. During the night we passed many interesting bits of scenery, and at different stations saw Arab sheikhs getting in and out of the train. Some of them were paid by the Turkish government to guard the line, and each saw the train through his own sphere of influence. Very few Christians come this way, but they are allowed to go up to Ma'an. When we were traveling, we only met one English-speaking tourist in the neighborhood of Amman, and he had the inevitable Cook's dragoman, a Syrian Christian dressed as a Muslim companion. We met them on our return again in the Austrian Lloyd Mail steamer, traveling from Beyrouth to Port Said. This satisfied us that they met with no bad luck in the country.

Amman, the ancient Rabbath Aman, was once besieged and taken by Joab; but afterwards belonged to the Ammonites. Ptolemy II (Philadelphus) of Egypt gave it the name of Philadelphia and it thrived for centuries belonging to the Decapolis. The ruins of Amman are the finest on the Eastern side of the Jordan, but are disappearing with the increase of the Circassian colony. The citadel commands a fine view of buildings of the Roman times, the remains of Thermae, the Street of Columns, and the Odeum, etc. etc. There is also a mosque of the time of the Abbasids and a fine tomb (Kabr-es-Sultan) with a triple vestibule, the central hall leading to a chamber with three shelf tombs.

Katrane-El-Hasna, Djirouf-Derwiseh, Aneze, through which we passed in succession are all historical places.

Ma'an is a big place, as things go in this part of the world. Someday there will be disturbances here like those of old Port Said. The ever-enterprising Greeks etc. have opened liquor shops here to the great disgust of old-fashioned Muslims. Indian beggars wandering on foot have also found their way here, intending to follow the kafila [caravan] from Syria, but as the Mahmal (the sacred covering or carpet) went by railway these unfortunate people were stranded. As they are accustomed to a roving life, they took it calmly. It was cold and they were badly dressed, food was dear and they were unprovided with money. What an ideal life! Oriental food is obtainable here, and we provide ourselves with some. At Amman, Tabuk, and at Ula one can also buy in stores and cook on the route. We followed the general custom. Postage stamps are sold here; but we were unable to procure any, although we explored the town, which was a short distance from the station. As the railway station was not provided with a bathing house and I had an inch of dust on me, owing to the journey, I tried a novel experiment of seeking the help of the engine driver, who gave me some hot water and with cold water from the tap in the engine shed. I had a glorious bath in the open air on a very cold day. Ma'an is the station beyond which only Muslims are allowed to travel without special permission. We started thence on a Sunday afternoon, towards the Betn-Ghel country. We passed Akabah, Mudeweri, Zat-ul-Haj, Tabuk, Jenain-ul Kazi, Almoazzam, Dae-el- Hamra, and Medain Saleh. The cities of the prophet Saleh, famous for his miracle of the "she-camel," need thorough exploration. The appearances of the surrounding country are awe-inspiring and grotesque. Some huge volcanic upheaval must have occurred here, overturning mountains, towns, and buildings. [Number XXXI, Jeremiah XLIX 28]. Kalat-uz-Zamurrud,

Hadish, the stables of Antar, Abu Jabir are all well-known places and the country gets somewhat fertile as we advance toward Medina.

An hour or so before the train reaches the holy city one begins to get a view of its buildings, the feeling is unique and thrilling, and every pilgrim recites the customary prayers. The military guards who accompany the railway for protection and travelers who have previously visited the city of peace take delight in pointing out the different objects of interest to the newcomers, whose exultation knows no bounds.

We reached Medina about 4:00 p.m. on a Tuesday and were met by kind friends, who accompanied us to the Bagh Shamsiah in two small carriages; but before going to our quarters we performed the asr and maghrib prayers at the Prophet's Mosque.[15] We appointed members of the well-known Birzanji family of S. Abdul Karim as our mozawwir cicerones.

After the Maghrib prayers I went through the regular ceremony of the first presentation. Owing to the arrival of the Mahmal from Syria the mosque was lighted up in the evening and kept open all night, which was a lucky thing for us. With the exception of a short time for meals and a little rest we spent that night and the greater part of the next few days in reciting prayers of various kinds. The zuhr and asr prayers are performed in the mosque at the earliest hour for each, and there is ample time between the various services, for the recitation of the Quran and vazifas. I had the supreme felicity of securing an excellent seat in the first row just behind the imam. There I attended for the five daily prayers, as well as the special Friday service and various extra-devotional prayers. At night the city gates were closed; but as early as 3:00 a.m. after our patiently waiting and cajoling of the keepers, by calling them fathers or brothers and impressing upon them that we wanted to attend the tahajjud prayers, they used to let us in. Then we had to wait at Bab-un-Nissa of the Prophet's mosque, which was the first gate opened on time to allow people to say the tahajjud prayers. We always stayed on for the morning prayers and had plenty of time for religious mediation and the reading of sacred formulae. I got enrolled among the various sets of attendants of the Prophet's tomb and mosque and paid my respects to the learned men. The great doctors of learning, Syed Muhammad Amin Rizwan and Syed Ahmed al-Barzanji, honored me by accepting me as their humble pupil and granted me diplomas for the reading of *ahadis* and *doas*. Maulana Moulvi Abdul Baki Sahib and Muhammad ibn Yousuf Malak Bashli of the family of Maolana Syed Ali Hurari also kindly granted me permission for recitation of the *Kasida Gousia* and *Dalail-ul-Khairat* respectively, within the precincts of the sacred *rauza*.

Medina (Lathrippa of Ptolemy, approximately 25° N, and 40° E), has long been a famous city. Baki, Yakut, Ibn Abd Rabah (tenth century), Ibn Jubair (1184), Ibn Khallikan (thirteenth century) have all written about it. Burckhard (1815) and Burton (1853) have given their views after observation and other Christian writers by hearsay. Since 1908 it has become the terminus of the Hijaz Railway, about 820 miles from Damascus. Medina is one of the best watered spots in Northern Arabia, owing to the convergence of torrent courses in its neighborhood. Wadi Buthan, a tributary of Wadi Kanat, passes through the suburbs of Medina and Wadi Akik is some miles to the South. The city, with its main paved street (seventh century); its massive circling walls by Jamal-ud-din of Isfahan (twelfth century); its rock fort and tiny buildings, has a population of about 20,000 souls. The green dome of the Prophet's tomb, his mosque (500 x 300 feet), and the date groves, which surround the town, would make any town picturesque; but its spiritual sanctity makes it a center for Muslims of the world, who can afford to disregard Burton's sneers. Muslim architecture is the admiration of the world, and it was possible for them to have surpassed themselves at Medina; but the great merit of the Prophet's last resting place is simplicity. Building or no building, the place is full of heavenly bliss for the true believers and the heart is nowhere so much at ease as in the Prophet's presence. From the religious point of view the primitive mosque, which was first built there, will never be surpassed: but all the subsequent constructions have been things of beauty and joy to the true believers.

Volcanic eruptions (latest in 1266), fires (in 1256, 1481), and sacks (last in 1804) have only helped to increase the worldly magnificence of the place, while detracting nothing from its sanctity or real greatness. The description given by Burton of the city and its buildings is pretty accurate and no great point can be made by correcting him in details; but we cannot help entering a protest against his unwarranted sarcasms and cynicisms. As a worshipper of the same God as Muhammad, Burton deals fairly with the impressiveness of Mecca; but he makes no attempt to hide his prejudice against the "Servant of God" and makes remark after remark without justification when dealing with Medina. It is fortunate that the Prophet can afford to disregard the gallant colonel's onslaughts. With all his knowledge of languages he misquotes Samanhudi, as explained by the editor of the *Encyclopedia Britannica*, and finds relief in relying upon other discredited authorities for his venomous remarks. Burton throws doubt upon the real tomb of Muhammad; but there is no reason to dispute the Muslim belief. There is no doubt about the grave. There is no proof of the Shias having ever removed the holy remains of the Prophet from Medina. Prophets are generally interred where they happen to die. It is said that Ali and the two sons of Abbas interred Muhammad close to the place where he had died and thence no earthly hand can remove him. Burton has discussed the position and the structure of the three tombs inside the Hujra; some of his remarks are based on weak traditions. There is no reason for attributing any mystery to them and the general belief of the Muslims is correct. There is no sepulcher, as distinct from the graves. The "light" associated with holy places and holy men is the spiritual Nur and not the sort of earthly light imagined by Burton. He finds a rational explanation for what he is pleased to call a Medinite Superstition "concerning the Column of Light which crowns the Prophet's tomb." He forgets that lamplight blazes in Medina only take place on fixed nights of the year and Orientals have keenness enough to discern between physical objects and heavenly ones. Spiritual light, or nur, is well-known and understood by professors of all revealed religions. Burton offends his own religious principles by scoffing at Islam. Burton evidently felt a pride in his successful disguise; but Europeans have an exaggerated idea of the difficulties of an adventure like his; many worse men have been to Mecca and Medina. As to the diploma of his murshid, it can be obtained from many printing houses, copyists and the so-called pirs, without getting oneself subjected to an examination of the inner man. They include a man in the category of Muslim, who openly pronounces the formula. There is a deeper significance in a Christian traveler having been permitted to see things. His writings have spread knowledge amongst Europeans, who can judge how noble, simple, and truthful are the practices of Islam.

Notes

1. He had worked as an attaché to the Earl of Northbrook during a mission in Egypt. See page 213 for further details.
2. Hamidullah had once served as his secretary in Egypt.
3. The couple met with him in London during this trip.
4. This reflection recalls his work with the British administration in Egypt alongside his father in the 1880s.
5. These "short excursions" apparently included London and Constantinople, since these dates encapsulate his entire trip through the Middle East and Europe!
6. Here Hamidullah seems at pains to emphasize Muslim unity before a European readership.
7. Burton, *Personal Narrative.*
8 *The Leader,* October , 1913, 6.
9. *The Leader,* October 30, 1913, 6.
10. *The Leader,* November 28, 1913, 6.

11. *The Leader*, January 31, 1914, 6.

12. Both these terms refer to types of Ottoman administrative districts.

13. Begum Sarbuland notes that that he had bought clothing for their servant Amina bi here. He, however, does not mention this.

14. A *kaymakam* was a midlevel Ottoman governor.

15. This line suggests that Hamidullah's use of "we" is the royal singular, for his wife did not in fact join him in these prayers, as indicated in her diary.

GLOSSARY

Allahu akbar An Arabic phrase meaning "God is great."

angrez This term may refer to a European, an English person, or a Westerner more generally, depending on context.

angrezi The adjectival form of *angrez*: English, European, Western.

apa "Elder sister," a reverential term that may be used literally or figuratively.

ashrafi A type of gold coin.

asr The afternoon prayer.

azan The Muslim call to prayer.

bahadur When placed after a name, a suffix that denotes honor or respect. It literally means "brave."

begum Historically, a royal or aristocratic title in Central Asia. It is now widely used to denote any Muslim woman of high rank, particularly in South Asia. It may also denote a married woman.

bhabhi Brother's wife, sister-in-law.

bhai Literally "brother" but also used as a term of respect. Often appended to the end of a first name to denote a close relationship.

bibi A respectful term that refers to upper-class or well-born women, often Muslim. It generally corresponds to the English word "lady."

bismillah A common Arabic phrase that means "in the name of God." It is typically read at the start of an activity

burqa One of many types of outer garments worn by women that covers both the body and face. Its style and coverage vary.

dastar-khwan A cloth spread on the ground on which a meal is served or eaten.

durud Prayers or blessings said for the Prophet Muhammad.

fajr The morning prayer, to be completed before dawn.

fatiha The fatiha is the first chapter of the Qur'an. It is commonly recited over graves.

al-hamdulillah "Praise be to God." An Arabic phrase that is used widely by Muslims of all backgrounds.

haram An Arabic word that literally means "sanctuary" or "holy space." When combined with the word *sharif* (noble) and used as a proper noun, it typically refers to the precincts of the main mosque in Mecca or the mosque in Medina that contains the tomb of the Prophet.

Hatim Also known as the Hajr Isma'il, this is a small wall opposite the Kaaba that lies within the circular space around which pilgrims walk while performing the tawaf, or ritual circumambulation. The term also refers to the area between the wall and the Kaaba itself.

hazrat A title of respect. When combined with a name, it refers to deceased saintly individuals particularly.

ihram The clothing worn while performing the hajj or umrah pilgrimages. Men typically wear two white, unstitched cloths. Women's clothing may vary but often does not include a veil or face covering.

imam zamin This is a small amount of money dedicated to a protecting figure, or imam, inserted into a band and tied to the arm of a departing person to protect them during their journey.

inshallah Literally, "if God wills it." A phrase applied to any future plan or to denote uncertainty.

isha The late evening prayer.

labbaik Roughly meaning "I am present" or "I heed your call" in Arabic, this word expresses one's intention to perform the hajj or umrah pilgrimages only for God's glory. It is repeated an unlimited number of times before and during these pilgrimages.

maghrib The dusk prayer.

marhaba An Arabic greeting, roughly meaning "welcome."

masha Allah "God has willed it." A widely used Arabic phrase that conveys a range of meanings, including happiness and thankfulness. It is the past tense form of *inshallah*, "if God wills it."

mem A shortened form of *memsahib*.

memsahib/memsahiba A South Asian term used to refer to Western, particularly British English, women. It derives from the word "madame." Begum Sarbuland also uses it to refer to Christians of any background.

miyan Young man.

muallim A guide that leads pilgrims through various ritual actions and prayers, who might also be tasked with arranging accommodation and transportation.

nawab Depending on era and context within Mughal and colonial India, an independent king, a subordinate ruler or governor, or an aristocrat, typically Muslim.

nafl A category of optional prayers that may be performed during or after certain activities.

niqab A type of face covering or mask worn by some Muslim women.

pir A title for a Sufi spiritual guide.

rakat A term that refers to one set, or cycle, of various prescribed movements and actions performed during prayer.

sahib (1) An honorific, typically affixed to the end of a name. (2) When used by women, it may be used to refer to one's husband. (3) A respectful term used to refer to a man, often a European or Western man. Begum Sarbuland also uses it to refer to Christians.

salaam Literally, a greeting or expression of peace. Salaams can be directed toward sacred spaces, such as the Kaaba or the grave of a saint.

sarkar An honorific title. Begum Sarbuland uses it to refer to her husband.

shaghdaf A camel litter; an enclosed space mounted on a saddle to allow the occupant, typically women, to travel unobserved.

sharif (1) An honorific placed after the name of a place; (2) an adjective used in South Asia to indicate a person of noble or respectable background.

Surah A chapter of the Qur'an.

tabarruk Items that have been blessed, typically by their presence in a sacred space or event.

tahajjud Supererogatory prayers performed during the late night.

tahlil To praise God by reciting the phrase *La ilaha ila Allah*, "there is no god but God."

takbir To recite the phrase *Allahu akbar*, "God is great."

tasbih Literally "to recite '*subhan Allah*,'" or "glory to God." It also refers to an object used to count these recitations, typically a looped string of beads that is used like a rosary.

tawaf The act of circumambulation, typically of the Kaaba.

umrah Referring to the "minor" or "lesser" pilgrimage, it involves circumambulation of the Kaaba and may be completed at any time of the year. It contrasts with the "major" pilgrimage, the hajj, which may only be completed at a specified time and which requires several more rituals.

vazifa A litany that typically includes Qur'anic verses, supplications, and other prayers.

vuzu Ritual ablution, or the act of washing to purify oneself before praying or handling the Qur'an, typically according to a set order.

Zamzam A revered well in Mecca; may also refer to water sourced from the well.

zuhr The late morning prayer.

BIBLIOGRAPHY

Known travel accounts published by Nawab Sarbuland Jung

In English

Diary of an Attaché in Egypt. Hyderabad, 1910.
A Pilgrimage to Mecca and the Near East. Secunderabad: Bulletin, 1912.
A Pilgrimage to Medina. Allahabad: Anwar-e Ahmadi, 1914.
"Floating Along II." *The Leader* (Allahabad), July 26, 1913.
"Floating Along III." *The Leader* (Allahabad), August 3, 1913.
"Floating Along IV." *The Leader* (Allahabad), August 20, 1913.
"Floating Along V." *The Leader* (Allahabad), September 11, 1913.
"A Pilgrimage to Medina I." *The Leader* (Allahabad), October 2, 1913.
"A Pilgrimage to Medina II." *The Leader* (Allahabad), October 30, 1913.
"A Pilgrimage to Medina III." *The Leader* (Allahabad), November 28, 1913.
"A Pilgrimage to Medina [IV?]." *The Leader* (Allahabad), January 31, 1914.
The Early Life of the First Student of the MAO College. Edited by Mahomed Ullah and Abdul Majeed Khwaja. Allahabad: Allahabad Law Journal Press, 1916.
Leaves from the Life of al-Haj Afzal ul-Ulema Nawab Sarbuland Jung Bahadur. Edited by Mahomed Ullah. Allahabad: Pioneer, 1926.

In Urdu

Safarnama-e Madina Munavvara. Allahabad: Anwar-e Ahmadi, 1914.
Safarnama-e Qustuntuniya. Hyderabad: Qa'im, n.d. [1911–12?].

Newspapers Consulted

Amrita Bazar Patrika
The Leader
The Pioneer
Tampa Morning Tribune
Times of India
The Graphic

Books and Articles Cited

A Short Summary of the Proceedings of the First Session of the All-India Ladies' Association Held in Bhopal from 26th March to 29th March 1918 under the Presidentship of HH Nawab Sultan Jahan Begum, . . . Ruler of Bhopal. Bhopal: Sultani, 1918.

Agha Mirza Beg. *Karnama-e Sarvari, Ya'ni, Savanih-e Khudnavisht*. Aligarh: Matba'-e Muslim University, 1933.

Akhtar, Jamil. *Urdu meñ Jara'id-e Nisvan ki Tarikh*. Vol. 1. Delhi: Kitabi Dunya, 2016.

Akhtar, Nazia. *Bibi's Room: Hyderabadi Women and Twentieth-Century Urdu Prose*. Hyderabad: Orient BlackSwan, 2022.

Alam, Asiya. *Women, Islam, and Familial Intimacy in Colonial South Asia*. Leiden, Netherlands: Brill, 2021.

Aydin, Cemil. *The Idea of the Muslim World: A Global Intellectual History*. Cambridge, MA: Harvard University Press, 2019.

Begum In'am Habibullah. *Ta'ssurat-e Safar-e Yurap*. Lucknow: N.P., 1937.

Begum Sarbuland Jang. *Dunya 'Aurat ki Nazar meñ: Mashriq o Maghrib ka ek Safarnama*. Delhi: Barqi , 1935.

Beverley, Eric. *Hyderabad, British India, and the World: Muslim Networks and Minor Sovereignty, c. 1850–1950*. Cambridge: Cambridge University Press, 2015.

Bidwell, Robin. *British Documents on Foreign Affairs: Reports and Papers from the Foreign Office*. Vol. 12. Frederick, MD: University Publications of America, 1986.

Boyk, David. "Fatima Begum." In *Three Centuries of Travel Writing by Muslim Women*, edited by Siobhan Lambert Hurley, Daniel Majchrowicz, and Sunil Sharma, 158–67. Bloomington: Indiana University Press, 2022.

———. "Provincial Urbanity: Intellectuals and Public Life in Patna, 1890–1930." PhD diss. University of California, Berkeley, 2015.

Burton, Richard Francis. *Personal Narrative of a Pilgrimage to Al-Madinah and Mecca*. New York: Dover, 1964.

Campbell, A. Claude. *Glimpses of the Nizam's Dominions*. Philadelphia: Historical Publishing Company, 1898.

Civil and Military Gazette. "India House Reception: The Guests." June 13, 1937.

Cohen, Benjamin. *An Appeal to the Ladies of Hyderabad: Scandal in the Raj*. Cambridge, MA: Harvard University Press, 2019.

Coronation Durbar-Delhi 1911: Official Directory with Maps. Calcutta: Superintendent Government Printing, 1911.

Dalrymple, William. *White Mughals: Love and Betrayal in Eighteenth-Century India*. London: Harper Collins, 2012.

Danapuri, Shah Muhammad. *Sair-e Dihli*. Agra: Riyaz-e Hind, 1894.

Datla, Kavita. *The Language of Secular Islam: Urdu Nationalism and Colonial India*. Honolulu: University of Hawai'i Press, 2013.

Datta, S. E. "Towards Real Comradeship." *Indian Magazine and Review*, December 1914.
Fatima Begum. *Hajj-e Baitullah o Ziyarat-e Diyar-e Habib*. Lahore: Kutub Khana Paisa Akhbar, 1959.
Fisher, Michael. *Counterflows to Colonialism: Indian Travellers and Settlers in Britain, 1600–1857*. Delhi: Permanent Black, 2004.
———. *Indirect Rule in India: Residents and the Residency System, 1764–1858*. Delhi: Oxford University Press, 1991.
Fyzee-Rahamin, Atiya Begum. *Atiya's Journeys: A Muslim Woman from Colonial Bombay to Edwardian Britain*. Edited by Siobhan Lambert-Hurley and Sunil Sharma. New Delhi: Oxford University Press, 2010.
Green, Nile. "Spacetime and the Muslim Journey West: Industrial Communications in the Making of the 'Muslim World.'" *American Historical Review* 118, no. 2 (2013): 401–29.
Hyderabad State: List of Leading Officials, Nobles and Personages. Hyderabad: Hyderabad Residency Government, 1937.
In Memoriam. Bombay: Printed at the Times Press, 1908.
Iqbal Ali, Sayyid. *Sayyid Ahmad Khan ka Safarnama-e Panjab*. Aligarh: Aligarh Institute, 1884.
Junne, George. *The Black Eunuchs of the Ottoman Empire: Networks of Power in the Court of the Sultan*. London: I. B. Tauris, 2016.
Keen, Caroline. *Princely India and the British: Political Development and the Operation of Empire*. London: I. B. Tauris, 2012.
Khan, M. Hameed-Ullah. *The Life and Times of Hafiz of Shiraz*. Allahabad, 1892.
Khan, Mehdi Hasan Khan. *Maharaja Kishen Parshad ki Zindagi ke Halat*. Hyderabad: Dar ul-Matba'-e Hukumat-e Hyderabad, 1951.
Khan, Sami'ullah. *Musafiran-e Landan*. Edited by Asghar 'Abbas. Aligarh: Educational Book House, 2012.
———. *Safarnama-e Maulvi Muhammad Sami'ullah Khan*. Amroha: 'Umdat ul-Matabi', 1880.
Khwaja, Jamal. *Vision of an Unknown Indian*. Lahore: Al-Hamd, 2011.
Laiq Ali Khan, Mir, Sunil Sharma, and Omar Khalidi. *Vaqayi'-e musafarat-e Navab . . . Mir Layiq 'Ali Khan*. Tehran: Nashr-e Tarikh-e Iran, 2009.
Lambert-Hurley, Siobhan. "Contesting Seclusion: The Political Emergence of Muslim Women in Bhopal, 1901–1930." PhD diss. SOAS University of London, 1998.
———. *Elusive Lives: Gender, Autobiography, and the Self in Muslim South Asia*. Stanford, CA: Stanford University Press, 2018.
———. "Out of India: The Journeys of the Begam of Bhopal, 1901–1930." *Women's Studies International Forum* 21, no. 3 (May 1998): 263–76.
Lambert-Hurley, Siobhan, Daniel Majchrowicz, and Sunil Sharma, eds. *Three Centuries of Travel Writing by Muslim Women*. Bloomington: Indiana University Press, 2022.
Lelyveld, David. *Aligarh's First Generation: Muslim Solidarity in British India*. New York: Oxford University Press, 2003.
Leonard, Karen. "Hyderabad: The Mulki-Non-Mulki Conflict." In *People, Princes, and Paramount Power: Society and Politics in the Indian Princely States*, edited by Robin Jeffrey, 65–106. Delhi: Oxford University Press, 1978.
Majchrowicz, Daniel. "Begum Hasrat Mohani's Journey to Iraq." In *Worlds of Knowledge in Women's Travel Writing*, edited by James Uden, 140–67. Boston: Ilex Foundation, 2022.

———. "'Can't Touch This': Early Indian Muslim Responses to the Saudi Conquest of the Hijaz." *Journal of Urdu Studies* 3, no. 1 (2023): 63–87.
———. "Malika Begum's Mehfil: Retrieving the Lost Legacy of Women's Travel Writing in Urdu." *South Asia: Journal of South Asian Studies* 43, no. 5 (2020): 860–878.
———. "Rahil Begum Shervaniya: Life Aboard a Pilgrim Ship." In *Three Centuries of Travel Writing by Muslim Women*, edited by Siobhan Lambert-Hurley, Daniel Majchrowicz, and Sunil Sharma, 126–136. Bloomington: Indiana University Press, 2022.
———. "Sughra Humayun Mirza: Meeting the Caliph in Switzerland." In *Three Centuries of Travel Writing by Muslim Women*, edited by Siobhan Lambert-Hurley, Daniel Majchrowicz, and Sunil Sharma, 451–460. Bloomington: Indiana University Press, 2022.
———. *The World in Words: Travel Writing and the Global Imagination in Muslim South Asia*. Cambridge: Cambridge University Press, 2023.
Maulvi Muhammad Zaka'ullah. *Savanih 'Umari Haji Sami'ullah Khan Bahadur C. M. G.* Lucknow: Uttar Pradesh Urdu Academy, 1997.
Minault, Gail. "'I Am a Ruby Wrapped in a Bag': Zay Khay Sheen and the Possibility of Poetry as Autobiography." In *Urdu and Indo-Persian Thought, Poetics, and Belles Lettres*, edited by Alireza Korangy, 179–91. Brill's Indological Library, vol. 51. Leiden, Netherlands: Brill, 2017.
———. *The Khilafat Movement: Religious Symbolism and Political Mobilization in India*. New York: Columbia University Press, 1982.
———. "Master Ramchandra of Delhi College: Teacher, Journalist, and Cultural Intermediary." *Annual of Urdu Studies* 18 (2003): 95–104.
———. *Secluded Scholars: Women's Education and Muslim Social Reform in Colonial India*. Delhi: Oxford University Press, 1998.
Mirza, Sughra Humayun. *Roznamchah-e Safar-e Bhopal aur Agra aur Dilli ke Halat*. Hyderabad: Nizam-e Dakkan, n.d.
———. *Safarnama-e Yurap*. Vol. 2. Hyderabad: A'zam , 1926.
Muhammad Akbar Ali Beg. *Muhammad Aziz Mirza: Shakhsiyat, Hayat, aur Karname*. Hyderabad: Idarah-e Sher o Hikmat, 1987.
Muhammadi, Begum, and Kulsoom Husein. *A Long Way from Hyderabad: Diary of a Young Muslim Woman in 1930s Britain*. Translated by Zehra Ahmad and Zainab Masud. Delhi: Primus Books, 2022.
Naim, C. M. *Urdu Texts and Contexts: The Selected Essays of C. M. Naim*. Delhi: Permanent Black, 2004.
Nawaz, Sumaira. "East Is East and West Is West: An Account of Nazli Fyzee and Begum Sarbuland Jang's Travels to Europe, c. 1908–1910." Unpublished paper, 2020.
Oldenburg, Veena Talwar. *The Making of Colonial Lucknow, 1856–1877*. Princeton, NJ: Princeton University Press, 2016.
Peile, John. *Biographical Register of Christ's College, 1505–1905*. Vol. 2. Cambridge: Clarke, 1913.
Pernau, Margrit. "Female Voices: Women Writers in Hyderabad at the Beginning of the Twentieth Century." *Annual of Urdu Studies* 17 (2002): 36–54.
———. "Schools for Muslim Girls: A Colonial or an Indigenous Project? A Case Study of Hyderabad." *Oriente Moderno* 84, no. 1 (2004): 263–76.
Ramusack, Barbara. *The Indian Princes and Their States*. Cambridge: Cambridge University Press, 2004.
Robinson, Francis. *The `Ulama of Farangi Mahall and Islamic Culture in South Asia*. London: C. Hurst, 2001.

Server-ul-Mulk, Nawab Bahadur. *My Life: Being the Autobiography of Nawab Server-Ul-Mulk Bahadur.* Translated by Nawab Jivan Yar Jung. London: A. H. Stockwell, 1931.

Shervaniya, Rahil. *Zad us-Sabil, ya, Rahlat ur-Rahil.* Aligarh: Matbaʿ-e Muslim University, 1929.

Shibli Nuʿmani. *Safarnama-e Rum o Misr o Sham.* Delhi: Qaumi, 1901.

"Ship Fact Sheet—Salsette (1908)," n.d. https://www.poheritage.com/Upload/Mimsy/Media/factsheet/94429SALSETTE-1908pdf.pdf.

Sikandar Begum. *A Princess's Pilgrimage: Nawab Sikandar Begum's Pilgrimage to Mecca.* Edited by Siobhan Lambert-Hurley. Bloomington: Indiana University Press, 2008.

———. "Tarikh-e Safar-e Makka." Bhopal, 1867. Ref. number: PP MS 55, Durand, box 2, file 12. SOAS.

StoryCorps Archive. "Syed Family Interview, Zakariya Syed w/ Akbar Jung." Accessed April 18, 2022. https://archive.storycorps.org/interviews/syed-family-interview-zakariya-syed-w-akbar-jung/.

"The System of Administration in the Nizam's Domain." Amrita Bazar Patrika (Calcutta), July 31, 1908.

Tahsin, Aminah. *Haidarabad meñ Urdu ka Nisa'i Adab.* Delhi: Educational Publishing House, 2016.

Tampa Morning Tribune. "Elephants and Motorcars at Wedding." June 18, 1916, Magazine sec.

Times of India (1861–2010). "Social and Personal." August 20, 1925.

Times of India (1861–2010). "Social and Personal." June 11, 1926.

Times of India (1861–2010). "Social and Personal." February 10, 1932.

Ummat ul-Ghani, Nur un-Nisa. *Safarnama-e Hijaz, Sham, o Misr (1909).* Hyderabad: Wordmaster, 1996.

Withey, Lynne. *Grand Tours and Cook's Tours: A History of Leisure Travel, 1750 to 1915.* New York: Morrow, 1997.

Zamindar, Vazira Fazila-Yacoobali. *The Long Partition and the Making of Modern South Asia: Refugees, Boundaries, Histories.* New York: Columbia University Press, 2007.

INDEX

DANIEL MAJCHROWICZ is Associate Professor of South Asian Literature and Culture at Northwestern University. He is author of *The World in Words: Travel Writing and the Global Imagination in Muslim South Asia* and editor (with Siobhan Lambert-Hurley and Sunil Sharma) of *Three Centuries of Travel Writing by Muslim Women* (Indiana University Press, 2022).

For Indiana University Press

Lesley Bolton *Project Manager/Editor*

Tony Brewer *Artist and Book Designer*

Anna Garnai *Editorial Assistant*

Sophia Hebert *Assistant Acquisitions Editor*

Samantha Heffner *Marketing and Publicity Manager*

Brenna Hosman *Production Coordinator*

Katie Huggins *Production Manager*

Bethany Mowry *Acquisitions Editor*

Dan Pyle *Online Publishing Manager*